Lower Ordovician trilobites of the Kirtonryggen Formation, Spitsbergen

by

Richard A. Fortey and David L. Bruton

Acknowledgement

Financial support for the publication of this issue of
Fossils and Strata was provided by the Lethaia Foundation

Contents

Lower Ordovician trilobites of the Kirtonryggen Formation, Spitsbergen

RICHARD A. FORTEY AND DAVID L. BRUTON

Fortey, R. A. & Bruton, D. L. 2013: Lower Ordovician trilobites of the Kirtonryggen Formation, Spitsbergen. *Fossils and Strata*, No. 59, pp. 1–116. ISSN 0024–1164.

The Kirtonryggen Formation is a thick, Lower Ordovician, palaeotropical, shallow-water carbonate succession exposed in northern Ny Friesland, Spitsbergen, in outcrops adjacent to Hinlopen Strait. The trilobites from the earliest part of the Ordovician of Spitsbergen are described for the first time, based upon two collections made during Cambridge University (1967), and joint Norsk Polarinstitutt and Palaeontologisk Museum, University of Oslo expedition in 1972. This work completes the monographic treatment of the Ordovician trilobites of Spitsbergen, Svalbard archipelago. Previous research on the Ordovician of Svalbard is summarised, especially relating to the faunas of the overlying Valhallfonna Formation (Floian–Dariwillian) which represent diverse, deeper-water biofacies as compared with the faunas described herein. The Kirtonryggen Formation (Ibexian: Tremadocian–early Floian) is divided into three members: in ascending order, Spora, Bassisletta and Nordporten Members, each with distinct trilobites, which are nearly all Laurentian endemics belonging to the Bathyurid biofacies. The sequence is as complete as any on the eastern side of the Laurentian palaeocontinent. The Spora Member is Stairsian in age, with a small fauna dominated by hystricurids and leiostegiids (*Svalbardicurus delicatus* fauna). The Bassisletta Member is sparingly fossiliferous, but includes Tulean-age trilobites, including early representatives of Bathyuridae (*Peltabellia* to *Chapmanopyge* faunas). The Nordporten Member has a widespread Blackhillsian fauna in its upper part (*Petigurus nero* fauna) underlain by a distinctive, but related fauna (*Petigurus groenlandicus* fauna) in its lower part. The older fauna includes species in common with a fauna from Greenland described by Poulsen in 1937 and is late Tulean to earliest Blackhillsian in age. The Tremadocian–Floian boundary is placed late in the *Chapmanopyge* fauna. The Kirtonryggen Formation trilobites include species in common with many localities along the eastern margin of Ordovician Laurentia, and their sequential stacking in Spitsbergen has proved useful in establishing the stratigraphy elsewhere and confirms that individual species were widespread and of biostratigraphic utility. Correlations with Lower Ordovician trilobite faunas previously described from Canada, Greenland, western Newfoundland, Vermont–New York State, Oklahoma and Missouri are discussed. Faunas closely similar to those described from the Nordporten Member occur in Greenland and on the Northern Peninsula, western Newfoundland. In general, the Early Ordovician trilobite faunas in Spitsbergen have undoubted similarities with those known from eastern Laurentia, and less in common with those described from the Great Basin, western USA, the area which has become the stratigraphical standard for Laurentia. Restricted environmental conditions on the heterogeneous carbonate platform may have generated endemics in eastern Laurentia, compared with more open shelf conditions in western Laurentia (on present geography). Fifty-three species belonging to 31 genera are considered, of which 15 species are described as new, 15 identified with previously described taxa and 24 described under open or tentative nomenclature. Three new genera: *Svalbardicurus, Harlandaspis* and *Eurysymphysurina*, are proposed, and *Chapmanopyge* is introduced as a replacement name for *Chapmania* Loch, 2007, pre-occupied. Occurrence in Spitsbergen of almost all genera of Bathyuridae allows a review of some of the problems in classification of the family. Alphabetically by genus, new species are as follows: *Bathyurellus diclementsae, Bolbocephalus gunnari, Catochia hinlopensis, Eurysymphysurina spora, Illaenus (Parillaenus) primoticus, Jeffersonia striagena, J. viator, Lachnostoma platypyga, Leiostegium spongiosum, Peltabellia glabra, Phaseolops? bobowensi, Raymondaspis? pingpong, Svalbardicurus delicatus* and *Uromystrum drepanon*. Although shallow-water biofacies predominate, the appearance of asaphids, remopleuridids and shumardiids at the very top of the Kirtonryggen Formation indicates a short-lived deepening prior to the drastic facies change at the base of the Valhallfonna Formation, which is attributed to a foundering of the shelf. The lower Nordporten Member has yielded the earliest known occurrences of three major trilobite superfamilies: Illaenoidea, Proetoidea and Scutelluoidea, respectively, together with the oldest leperditicope 'ostracod'. This is consistent with hypotheses relating the origin of new major clades to inshore habitats. Subsequent Ordovician occurrences of these groups record their expansion on to different palaeocontinents and into deeper-water palaeoenvironments. □ *Biogeography, biostratigraphy, clade origin, Early Ordovician, Spitsbergen, Svalbard, taxonomy, trilobites.*

Richard A. Fortey [raf@nhm.ac.uk], Department of Earth Sciences, The Natural History Museum, London SW7 5BD, UK; David L. Bruton [d.l.bruton@nhm.uio.no], Natural History Museum (Geology), University of Oslo, Postboks 1172, Blindern NO-0318 Oslo, Norway; manuscript received on 28/08/2012; manuscript accepted on 20/05/2013.

DOI 10.1111/let.12029 © 2013 The Authors, Lethaia © 2013 The Lethaia Foundation

Introduction

The Ordovician succession adjacent to Hinlopen-stretet on northern Ny Friesland on the island of Spitsbergen, the largest of the Svalbard archipelago, is divided into two formations (Fortey & Bruton 1973). The younger of these formations, the Valhallfonna Formation, comprises mostly dark limestones and shales that were deposited in an outer shelf setting. The fossil fauna of the Valhallfonna Formation is remarkably rich, including trilobites, brachiopods, graptolites, conodonts, early vertebrates, radiolarians and molluscs. The admixture of these different elements makes the Valhallfonna Formation of considerable importance in stratigraphic correlation and biofacies recognition, in addition to the intrinsic interest of the well-preserved fossil fauna. Its age spans the Lower-to-Middle Ordovician boundary, the upper strata assigned to the Profilbekken Member being of entirely Whiterockian age, while the underlying Olenidsletta Member is mostly upper Ibexian (= Floian, or formerly Arenigian) in age, but contains the Ibexian–Whiterockian boundary in its upper part. The varied trilobite fauna of the Valhallfonna Formation has been monographed in considerable detail and was seminal in the recognition of trilobite biofacies (Fortey 1975b). The graptolites were described by Cooper & Fortey (1982) and include isolatable material, thus placing the biostratigraphy on a sound international footing. However, apart from a preliminary faunal list in Fortey & Bruton (1973), the trilobite fauna of the thick Kirtonryggen Formation underlying the Valhallfonna Formation has not received further attention. Correlative strata have been recognised recently on the other side of Hinlopenstretet, in Nordauslandet (Stouge *et al.* 2011), but detailed biostratigraphy remains to be done there. This work completes the account of the Ordovician trilobites of Spitsbergen, one of the richest localities known in rocks of this age.

Some of the species typical of the Kirtonryggen Formation were originally described from western Newfoundland by Billings (1865), and Fortey considered that it was essential to revise them before consideration of the Spitsbergen fauna. These trilobites, originating from the St George Group, were revised by Fortey (1979) and Boyce (1989). This has helped to provide a comparative standard for the taxonomy of the trilobites from the upper part of the Kirtonryggen Formation. Further revisions or additions to the Early Ordovician platform trilobite faunas of eastern Laurentia have been made subsequently, with descriptions of species from Okla-homa, New York State, eastern and arctic Canada and Greenland in recent years. Most of these are small faunas from various horizons within the Early Ordovician, Ibexian (Tremadocian–Floian). The particular importance of the Spitsbergen succession is that several different faunas are stacked one after the other in the Kirtonryggen Formation, which allows identification of the relative ages of fossiliferous horizons previously recognised elsewhere. For example, the faunas from Greenland described long ago by Poulsen (1937) can now be placed in their proper sequence.

This work describes the trilobites from northern Spitsbergen from the section through the Kirtonryggen Formation along Hinlopen Strait collected during the 1968 Cambridge University and 1972 Oslo Paleontological Museum field expeditions. The majority of the specimens were collected during the latter trip. The only fossils so far described from the Kirtonryggen Formation include the earliest known leperditicope 'ostracod' from the upper part of the Nordporten Member, named as *Trinesos akroria* by Williams & Siveter (2008), and a few widespread Laurentian articulated brachiopods from the Spora and Nordporten Members (Hansen & Holmer 2011), and some recently reported conodonts (Lehnert *et al.* 2013). The trilobite fauna demonstrates the radiation in Ordovician bathyurid trilobites and other species adapted to an extensive, shallow subtidal to peritidal carbonate shelf that typified a vast area of the ancient Laurentian palaeocontinent during the earlier Ordovician. Furthermore, the fauna proves that some trilobite species were very widespread, making them potent stratigraphical tools to be applied all along the eastern margin (present geography) of former Laurentia. It also reveals that there were some interesting differences between the trilobite faunas of eastern and western Laurentia, typified by the richly fossiliferous, and frequently silicified, sequences of the Great Basin in Utah, Nevada and Idaho (Ross 1951; Hintze 1953; Adrain *et al.* 2009). Apart from the rich variety of endemic bathyurids, the fauna is also of considerable taxonomic interest in yielding the oldest known illaenid, proetid and styginid trilobites. This in turn relates to hypotheses proposing the inshore origin of major clades (Jablonski 2005).

The fauna described here is rich in species, but several of these taxa remain incompletely understood and are described under open nomenclature. Further collecting and research on the faunas of the Kirtonryggen Formation would be worthwhile to characterise these species adequately, but no further expeditions to this remote region have happened over the last four decades.

Summary of previous research on the Ordovician of Spitsbergen

Since this paper completes the account of the Ordovician of northern Spitsbergen, it may be useful briefly to summarise previous work. Several of the earlier papers describing different aspects of the faunas are published in journals without wide distribution, and some of these are in danger of slipping out of sight. The general reference on the geology of Spitsbergen is Harland (1997), and W. B. Harland was responsible for introducing the first author to Spitsbergen in 1967 on one of the Cambridge University expeditions. *Harlandaspis* is named for him in this work.

Hallam (1958) reported the discovery of Ordovician rocks and fossils in Ny Friesland by the Cambridge University expedition; they were collected near the top of the thick Hecla Hoek Group (Proterozoic–Ordovician). The molluscs and other fragments were indicative of Early Ordovician age and probably of similar age to the trilobites described in this paper. These outcrops near the large glacier known as Oslobreen, south of our study area, were mapped in more detail by a Cambridge University team in the years following, and a more detailed account of the stratigraphy was given by Gobbett & Wilson (1960), who proposed the Kirtonryggen Formation for the Ordovician strata. Gobbett (*in* Gobbett & Wilson 1960) also described the first Early Ordovician trilobite from Ny Friesland, *Hystricurus wilsoni* Gobbett. Interestingly, it is not identical to any hystricurid herein and would probably now be referred to one of the Skullrockian genera erected by Adrain *et al.* (2003). Hence, it is older than the earliest fauna we have discovered from the Spora Member, which is Stairsian in age.

Discovery of the sections along Hinlopen Strait (Hinlopenstretet) occurred accidentally, when the 1966 Cambridge University expedition stopped to collect water from a melt stream, one that later came to be called 'Profilbekken'. G. Vallance was an undergraduate student on that expedition. Preliminary determination of several different trilobites by H. B. Whittington encouraged a focussed collection trip in the summer of 1967, with Vallance assisted by Richard A. Fortey (RAF), during which some of the specimens figured herein were collected. The great diversity and good preservation of the faunas were immediately apparent. An outline account of the regional stratigraphy was published by Vallance & Fortey (1968). RAF then studied a fraction of the faunas for a PhD thesis and published the first paper on the new collections based upon isolated material

of the graptolite *Pseudotrigonograptus* (Fortey 1971). In 1972, an expedition of the Palaeontological Museum, Oslo, and the Norsk Polarinstitutt, including G. Henningsmoen, RAF and David L. Bruton (DLB), made extensive new collections from a measured section, from which the majority of the type specimens are derived. Fortey & Bruton (1973) published a map (see also Hansen & Holmer 2011; Lehnert *et al.* 2013) and outlined the regional stratigraphy, establishing the lithostratigraphic units in use today. These units have since been identified in north Eastland (Nordaustlandet) on the opposite side of Hinlopen Strait, and correlation has been made with other sites along the Iapetus borders (Smith & Rasmussen 2008; Stouge *et al.* 2011, 2012).

The fauna of the Valhallfonna Formation was studied intensively over the next decade, RAF devoting much of his time to the task. Two members were recognised: the upper Profilbekken Member including a typical Whiterockian trilobite fauna of North American type and the lower Olenidsletta Member a unique mixture of alternating biofacies yielding deeper-water assemblages of 'Arenigian' age. Four major successive faunas termed 'V$_1$' to 'V$_4$' were summarised in a range chart in Fortey (1980a). The first trilobite to be described was the pelagic *Opipeuterella* (Fortey 1973), which is now known to be widespread. The other trilobites of the Valhallfonna Formation were described in three monographs (Fortey 1974, 1975a, 1980a). In particular, the Ordovician radiation of the Olenidae (Fortey 1974) was remarkable and remains unparalleled from collections elsewhere. The co-occurrence of suites of particular genera was clearly related to Ordovician palaeoenvironmental conditions, which were discriminated in the Hinlopen Strait sections for the first time. Fortey (1975b) described them as 'community types', although 'biofacies' is the preferred term for the same concept in modern literature. A sequence running through olenid, nileid and illaenid–cheirurid biofacies with decreasing depth have now been widely recognised elsewhere. The principle that deeper-water assemblages tend to be more independent of geographic barriers is also generally adopted. Dissolution of limestone for graptolites revealed the presence of well-preserved radiolarians, some of which have been described by Fortey & Holdsworth (1971), Holdsworth (1977), and Maletz & Bruton (2007, 2008), also chitinozoa (Bockelie 1980, 1981). The systematics of the graptolites themselves were published by Fortey (1971), Archer & Fortey (1974), and Cooper & Fortey (1982). Correlation with the Australasian graptolite standard was established in these works. Residues from acid preparation included a range of phosphatic fossils. Fragments of vertebrate bone (*Anatolepis*) were the oldest record of the phylum at that time (Bockelie & Fortey 1976; Bockelie *et al.* 1976). Minute larval 'shells' were attributed to very early growth stages of trilobites (Fortey & Morris 1978), while the enigmatic fossil *Janospira* Fortey & Whittaker, 1976, was subsequently regarded as an aberrant mollusc. Another mollusc was discovered from silicified material and became the type species of a new bivalve genus, *Tironucula* Morris & Fortey, 1976. Conodonts recovered from the residues showed the same relation to biofacies as did the trilobites, a relationship described by Fortey & Barnes (1977). Unfortunately, the systematics of the conodonts were never published from the original collections, which is an omission given the biostratigraphic significance of such a 'mixed province' locality. However, conodonts from samples collected from the Kirtonryggen and Valhallfonna formations have recently been described by Lehnert *et al.* (2013). Hansen & Holmer (2011) have described the brachiopod faunas of the Valhallfonna and Kirtonryggen formations, the former having by far the richer assemblage of deeper-water genera.

The stratigraphic significance of the succession of faunas in the Valhallfonna Formation was summarised by RAF (Fortey 1976, 1980b). The particularly rich succession spanning an interval close to the base of the Whiterockian Laurentian regional stage indicated that knowledge of faunas of macrofossils from the standard Ibexian–Whiterockian successions in the Great Basin, western USA, was incomplete (e.g. Hintze 1953), probably because a sequence boundary is present there. Proposal of a Valhallan Stage to accommodate this interval did not meet with much acceptance, and subsequent definition of the base of the Whiterockian (Ross *et al.* 1997) effectively drew its lower limit to incorporate the Valhallan equivalents in Spitsbergen, equivalent to the upper part of the former 'Arenigian' Series of Europe (see also Fortey & Droser 1996). In modern global terms, the top of the Floian Stage and the base of the Dapingian, and hence the Middle Ordovician, lie at or close to the base of V_3 in the upper part of the Olenidsletta Member.

More general studies of the trilobite faunas showed that the deeper-water genera of the olenid biofacies, and carbonate 'mound' faunas of the illaenid–cheirurid biofacies, tended to have much longer stratigraphical ranges than genera of the open shelf nileid biofacies, where the greatest biodiversity was also present (Fortey 1980c). A similar association with biofacies has been recognised for the brachiopods based on our collections and those made in

2008 by Hansen & Holmer (2010, 2011). The Nordporten Member also yielded the earliest record of a leperditicope arthropod (Williams & Siveter 2008). The biofacies profile was again used to demonstrate the effect of the palaeoenvironment on trilobite cuticle thickness (Fortey & Wilmot 1991), showing that deepest faunas had only thin cuticles, while inshore faunas included species with the thickest cuticles, but that 'thin-shelled' species could also be found in specific niches in shallow-water habitats. This conclusion is reinforced in the present work: within the inshore Bathyurid biofacies, *Petigurus* is remarkably robust, yet the same environment also supported 'thin-shelled' species belonging to such genera as *Licnocephala* (see Fig. 26). Regrettably, our rock collections have not been studied in detail by sedimentologists, but lithological descriptions of parts of our section have been published in Russian by Kosteva & Teben'kov (2006).

Stratigraphy of the Kirtonryggen Formation

Stratigraphic sub-divisions and age

The type section of the Kirtonryggen Formation lies near Oslobreen, south of the Hinlopen Strait outcrop (Gobbett & Wilson 1960; Harland 1997). The present study is based on the northern part of Ny Friesland, where the formation is exposed along the shores adjacent to Hinlopen Strait (Hinlopenstretet), on the western side of the strait opposite the island of Nordaustlandet (Fig. 1). The lithostratigraphy of the Kirtonryggen Formation in this area was outlined by Fortey & Bruton (1973), where preliminary faunal determinations were also given. The lithologies exposed in the Kirtonryggen Formation are overwhelmingly of the type described from shallow-water Ordovician carbonate platforms laid down in tropical palaeolatitudes along the eastern margin of Laurentia, which have long been recognised as constituting a single entity in Early Ordovician times (Swett & Smit, 1972; Swett 1981; James *et al.* 1989; Derby *et al.* 2012 for review). For example, contemporary sequences very similar to those of the Kirtonryggen Formation have been described from the St George Group, western Newfoundland, by Fortey (1979), Knight & James (1988) and Boyce (1989), and from Greenland by Cowie & Adams (1957). New faunal evidence is consistent with this geographical continuity, and biostratigraphical sub-divisions of the carbonate platforms developed elsewhere can be broadly applied to the Spitsbergen

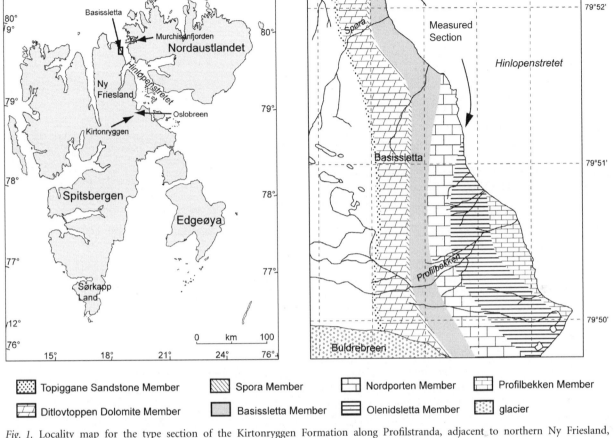

Fig. 1. Locality map for the type section of the Kirtonryggen Formation along Profilstranda, adjacent to northern Ny Friesland, Spitsbergen. After Fortey & Bruton (1973).

sequence. However, as noted below, there are faunal differences between the two sides of the Ordovician Laurentian palaeocontinent which make some difficulties in the straightforward application of the standard sub-divisions of the Ibexian Series detailed by Ross *et al.* (1982, 1997) based upon the succession of the Great Basin, western USA.

The discovery of productive horizons along the type section along Profilstranda was sporadic, the massive, rubbly weathering limestones were not easy to collect, and sections are hard to measure accurately. The considerable number of species that we have been obliged to leave in open nomenclature proves that further collections would extend the fauna. However, the sequence of faunas is well-established, and the range chart (Fig. 2) is related to the type section along Profilstranda. A few useful collections were made inland from isolated outcrops. These cannot be located precisely in the type section, but their approximate horizon can generally be determined with some confidence in relation to col-

lections made from the measured section. In the text and plate descriptions, these specimens are indicated by 'ca.' (circa) before the stratigraphic horizon appropriate to the main coastal section. It is also important to note that the two section measurements of the collections from the Oslo expedition and Cambridge University expedition do not mesh precisely. For the purposes of this paper, the Oslo section is taken as standard, as it was measured in more detail, and the Cambridge collections of illustrated material matched as closely to it as we can manage. This is not ideal, but is the product of different fieldwork conducted several years apart under different ice conditions.

As in northwest Scotland and East Greenland, strata of presumed middle and upper Cambrian age are as yet unproven from finely laminated dolomitic strata of the Tokammane Formation underlying the Kirtonryggen Formation, although there is no angular unconformity above the former. From faunal evidence given below, it is likely that the earliest

SM (20m)	Bassisletta Member (250m)				Nordporten Member (220m)		
SD	No fauna	P G	No fauna	*Chapmanopyge* fauna	*Petigurus groenlandicus* fauna	*Petigurus nero* fauna	LP

——*Svalbardicurus delicatus* n. gen. et sp.

——*Hystricurus* cf. *Hystricurus* sp. nov. B Adrain et al. 2003

——*Hystricurus* sp. 1

——*Leiostegium spongiosum* n. sp.

——*Eurysymphysurina spora* n. gen. et sp.

——*Pilekia* cf. *P. trio*

X *Peltabellia glabra* n. sp.

X *Chapmanopyge 'amplimarginiata'* (Billings)

Chapmanopyge cf. sp. 1 of Loch, 2007 ———

Jeffersonia viator n. sp. ——— – – – –

Punka latissima n. sp. ————————

Ceratopeltis cf. *C. batchensis* (Adrain & Westrop)   **X**

Chapmanopyge cf. *C. sanddoelaensis* Adrain & Westrop   **X**

Licnocephala n. sp. C ——

Carolinites? n. sp. A ————————

Petigurus groenlandicus (Poulsen) ————————

Phaseolops? *bobowensi* n. sp. ————

Bolbocephalus gunnari n. sp. ———— – – – –

Bathyurellus diclementsae n. sp. ————————

Uromystrum affine (Poulsen) **?**————————

Licnocephala n. sp. B **?**– – **?** – – – – **?**

Bolbocephalus sp. cf. *B. kindlei* Boyce **?**

Licnocephala n. sp. A **?**

Ischyrotoma parallela Boyce **?**– – – – – – –

Raymondaspis? *pingpong* n. sp. **?**– – – – – – –

Licnocephala brevicauda (Poulsen) **?**– – **X** – – – **?**

Jeffersonia striagena n. sp. ————

Illaenus primoticus n. sp. **X**

Randaynia n. sp. A **X**

Psalikilopsis aff. *cuspicaudata* Ross ——

Harlandaspis elongata n. sp. ——

Bolbocephalus convexus (Billings) ——————

Bolbocephalus stclairi Cullison ————

Catochia hinlopensis n. sp. ————

Chapmanopyge n. sp. A **X** – – – – – – – **?**

Jeffersonia timon (Billings) – – ——

Uromystrum drepanon n. sp. ———— – –

Bathyurine n. gen. et sp. A **X**

'*Jeffersonia*' aff. *J. granosa* Cullison **X**

Uromystrum aff. *U. affine* (Billings) ————

Catochia ornata Fortey ————

Bathyurellus abruptus Billings ————

Punka flabelliformis Fortey ————

Petigurus nero (Billings) ————

Benthamaspis conica Fortey ————

Licnocephala aff. *L. brevicauda* Poulsen **X**

Grinnellaspis newfoundlandensis Boyce **X**

Benthamaspis gibberula (Billings) ————

Ischyrotoma n. sp. A ——

Lachnostoma platypyga n. sp. **X**

Stenorhachis n. sp. A **X**

Eorobergia n. sp. A **X**

Conophrys sp. 1 **X**

Ordovician (Tremadocian, Skullrockian) may also be unrepresented by fossiliferous strata in northern Ny Friesland. Since Boyce (1989), Loch (2007) and Adrain *et al.* (2009) have all proposed trilobite-based faunal zones spanning the same interval in Early Ordovician Laurentian strata, there is no reason to add to a burgeoning list of biozones, and an informal terminology is applied in this work. Nonetheless, we can suggest a broad correlation between these different schemes (Fig. 3). The Kirtonryggen Formation was divided by Fortey & Bruton (1973) into three members on lithological grounds, which are also faunally distinct, as summarised by the species ranges in Figure 2. In order from oldest to youngest, these members are as follows:

Spora Member

Rather massive grey limestones, 20 m thick, are fossiliferous from the base of this unit to its top and yield the brachiopod *Syntrophina* along with numerous trilobites. With its rubbly occurrence, it has not proved possible to sub-divide this member, and the fauna appears to be the same throughout. The trilobites are compared in detail in the systematic section, but include the common hystricurid *Svalbardicurus delicatus* n. gen., n. sp., with rarer *Leiostegium spongiosum* n. sp., and *Eurymphysurina spora* n. gen., n. sp. and *Hystricurus* spp. *Svalbardicurus delicatus* is very similar to a species from the Lower Member of the Boat Harbour Formation in western Newfoundland, and we also compared *Leiostegium spongiosum* n. sp. to a species from the same unit (Boyce 1989). The brachiopod *Syntrophina* is commonly recorded in lower Ibexian strata (e.g. Hintze 1953). *Leiostegium* appears in the *Kainella/Leiostegium* (Zone D) Biozone of the standard Ibexian sections in western USA (Ross *et al.* 1997, p. 17), which is the earliest zone of the Stairsian regional stage. We compare a *Pilekia* species closely with *P. trio* Hintze, a species known from 'Zone E' in Utah. Ross *et al.* (1997) stated that the Stairsian interval lies above the lowermost Ordovician regional stage, the Skullrockian, which includes at least three Ordovician trilobite zones apparently not represented in either western Newfoundland or Ny Friesland. The Spora Member fauna appeared suddenly and endured briefly, and is likely early, but not earliest Ibexian (Stairsian, early Tremadocian). Thus, the disconformity beneath the base of the Kirtonryggen Formation also embraces the earliest Ordovician. Ross *et al.* (1997) noted that there is a 'low-diversity interval' among conodont faunas through the equivalent of the greater part of the *Kainella/Leiostegium* trilobite Biozone, and a similar and correlative interval has been identified in the St George Group in western Newfoundland and elsewhere by Ji & Barnes (1994). Lehnert *et al.* (2013) record conodonts of the *Rossodus manitouensis* Zone from this unit.

In the text, we refer to the lower fauna as the *Svalbardicurus delicatus* fauna. The fauna was apparently uniform throughout a massive, rubbly limestone, and the field collections were not sub-divided for this unit.

Bassisletta Member

The 250-m-thick Bassisletta Member is dominated by dolomites, oolites, stromatolites, edgewise conglomerates and other lithologies providing evidence of supratidal, intertidal or very shallow sub-tidal conditions. Not surprisingly, it is difficult to find macrofossils, trilobites included. One horizon 60 m from the base, a limestone between stromatolite 'heads', has yielded well-preserved material of *Peltabellia glabra* n. sp. This species is exceedingly like a species from the Barbace Cove Member of the Boat Harbour Formation that Boyce (1989, p. 150) named as *Peltabellia* sp. cf. *P. peltabella* (Ross). This was in turn placed in the synonymy of *Strigenalis implexa* Loch, 2007 by Loch (2007, p. 47) (see taxonomic discussion below), a species ranging through his *Benthamaspis rhochmotis* and *Petigurus cullisoni* biozones of the Kindblade Formation, Oklahoma. Boyce correlates the Barbace Cove Member with the Tulean Stage of the Ibexian Series, and this is not inconsistent with the assignment in Oklahoma, although Loch points out the differences between the local successions there and in the Nevada/Utah standard. The trilobite occurrence in the Bassisletta Member below the typical later Ibexian faunas of the Nordporten Member is consistent with an early Tulean age for this part of the Bassisletta Member.

Fig. 2. Stratigraphic distribution of trilobite species described in this paper through the three members of the Kirtonryggen Formation. SM, Spora Member; SD, *Svalbardicurus delicatus* fauna; PG, *Peltabellia glabra* fauna; LP, *Lachnostoma platypyga* fauna (top 3 m of Nordporten Member). Specimens found in the Spora Member are indicated as passing through the entire member, which was not subdivided in the original collections. X indicates occurrences from a single horizon. ? indicates specimens from the H collection (p. 9), low in *P. groenlandicus* fauna, exact horizon unknown and tentative extensions of ranges. Solid lines represent certain stratigraphic distribution, and dotted lines portray possible range extensions.

SPORA	BASSISLETTA		NORDPORTEN			Member (Not to scale)
TREMADOCIAN			FLOIAN			Global Stages
Svalbardicurus delicatus	? Peltabellia glabra	? Chapmanopyge	Petigurus groenlandicus	Petigurus nero	Lach. plat..	Kirtonryggen 'Faunas'
Randaynia saundersi	? Missing		? Strigigenalis brevicaudata	Strigigenalis caudata		Western Newfoundland Zones
?	R. brevicephalus - B. rhochmotis	P. cullisoni - B. stitti	Strigigenalis caudata			Oklahoma Zones
Leiostegium to Tesselacauda	Litzicurus shawi through	Psalikilus pikum	S. plicalabeona to C. nevadensis			Utah/Nevada
STAIRSIAN	? TULEAN		? BLACKHILLSIAN			Ibexian 'stages'

Fig. 3. Suggested correlation of successive Spitsbergen trilobite faunas with zonal schemes suggested elsewhere in Early Ordovician Laurentia. Not to scale.

In the text below, we refer to this fauna as the *Peltabellia glabra* fauna.

The upper part of the Bassisletta Member includes two trilobite faunas: the lower one is incompletely known, but includes *Chapmanopyge 'amplimarginiata'* (Billings). An upper fauna is better characterised, including a fauna with *Chapmanopyge* cf. sp. 1 (of Loch, 2007), *Punka latissima* n. sp., *Jeffersonia viator* n. sp., *Ceratopeltis* cf. *C. batchiensis* Adrain and Westrop, *Carolinites?* n. sp. A, and a *Chapmanopyge* sp. compared with *C. sanddoelaensis* (Adrain and Westrop). Comparison with the previously described species of these genera detailed in the systematic section suggests that this part of the member is assignable to the upper part of the Tulean Stage. Detailed comparison of *Jeffersonia viator* n. sp. indicates that it may be identical with a species described from the 'Jeffersonian' of Missouri and possibly one from the Baumann Fiord Formation of Ellesmere Island. This interval in the section is under-represented in the collections and would repay further exploration, which is reflected in a number of additional interesting species recorded in open nomenclature.

This collection of faunas is referred to the *Chapmanopyge* fauna in the systematic part. Adrain *et al.* (2009) have finely divided this part of the Early Ordovician into a number of biozones based upon silicified faunas in the Great Basin. Such refinement is hardly possible from our collections. Lehnert *et al.* (2013) record conodonts of the *Macerodus dianae* Zone from the upper part of the Bassisletta Member. This is regarded as upper

Tremadocian. However, it is likely that the uppermost part of the *Chapmanopyge* fauna is Floian in age, as concluded from conodonts obtained from samples processed by Dr M. P. Smith and reported below.

Nordporten Member

As already noted by Fortey & Bruton (1973), the 220-m-thick Nordporten Member is rich in fossils compared with the earlier part of the Ibexian succession. Trilobites are invariably disarticulated and sorted into different sclerites, which has presented particular problems in assigning all the exoskeletal parts to some taxa.

The Nordporten Member fauna divides into several successive intervals typified by different assemblages of bathyurid and other trilobites.

1 Lowest fauna, at the very base, intermediate with the younger *Chapmanopyge* fauna with species range extensions upwards of species from the top of the Bassisletta Member. This interval requires further fieldwork for clarification, since the collections may have been 'lumped'.

2 Diverse bathyurid dominated fauna with *Petigurus groenlandicus* Poulsen, 1937, *Uromystrum affine* (Poulsen, 1937), *Harlandaspis elongata* n. gen., n. sp. *Bathyurellus diclementsae* n. sp., *Licnocephala brevicauda* (Poulsen, 1937), *Illaenus primoticus* n. sp., *Raymondaspis?* pingpong, *Phaseolops?* bobowensi n. sp., *Bolbocephalus gunnari* n. sp. *Psalikilopsis* n. sp., and particularly

known from abundant material 105 m from base of member (115 m from top), but extending from the base to 130 m. We refer to this assemblage as the *Petigurus groenlandicus* fauna below. One collection made by Gunnar Henningsmoen inland away from the type section is probably from the same interval, but includes some species that have not been found elsewhere, and lies within the lower part of the *P. groenlandicus* interval. It is referred to as the 'H collection'.

The fauna of the lower part of the Nordporten Member is termed the *Petigurus groenlandicus* fauna below

3   Diverse bathyurid dominated fauna with *Petigurus nero* (Billings), *Uromystrum drepanon* n. sp., *Grinnellaspis newfoundlandensis* Boyce, *Bathyurellus abruptus* Billings, *Catochia hinlopensis* n. sp., *Punka flabelliformis* Fortey, *Bolbocephalus convexus* (Billings) *Benthamaspis gibberula, B. conica* and others. Many of these species were originally named by Billings (1865) from the Catoche Formation of western Newfoundland, a fauna revised by Fortey (1979) and Boyce (1989). This is the classical 'Upper Canadian' (or Cassinian) fauna of the older literature. It is evidently widespread, as it is found additionally in the Wandel Valley Formation, eastern North Greenland (Fortey 1986) and in the Fort Cassin Formation of Vermont (Brett & Westrop 1996). There is no reason to revise the assessment of its age originally given by Fortey (1979) as corresponding with the *Trigonocerca typica* Biozone (Zone H) of the Great Basin standard, which is now referred to the early Blackhillsian regional stage of the Ibexian (Ross *et al.* 1997, p. 19; Adrain *et al.* 2009). This fauna is present through an interval from 130 to 200 m of the Nordporten Member.

This is referred to as the *Petigurus nero* fauna below.

4 The uppermost 3 m of section comprise more fissile and darker limestones than those below. There is a radical change in faunal composition with the appearance of abundant asaphids attributed to *Lachnostoma platypyga* n. sp. and *Stenorhachis* n. sp. A in the uppermost Kirtonryggen Formation, together with *Conophrys* and a remopleuridid; with the exception of *Benthamaspis*, bathyurids disappear. This is clearly a facies change from the faunas dominated by bathyurid trilobite species underlying it, and we consider that the fauna may not be significantly younger

than the *Petigurus nero* fauna. As discussed below, the interval probably represents a deepening event transitional to the conditions pertaining during the deposition of the Valhallfonna Formation and may be stratigraphically condensed.

This is referred to as the *Lachnostoma platypyga* fauna in the systematic account.

Lehnert *et al.* (2013) reported on conodonts from the Nordporten Member. An assemblage indicating the *Oneotodus costatus* Zone was obtained from the uppermost Bassisletta Member and the lower part of the Nordporten Member (*Petigurus groenlandicus* fauna). They report the transition to the *Oepikodus communis* Zone within the upper part of the Nordporten Member (their sample BS 23), which we cannot locate precisely in our section. Since the same conodont fauna continues into the Valhallfonna Formation, however, this does support the interpretation of a facies change rather than a hiatus at the junction of the Kirtonryggen and Valhallfonna formations.

Tremadocian–Floian boundary

The Kirtonryggen Formation is overlain conformably by the Valhallfonna Formation, at the base of which there is a drastic change in lithology and biofacies indicating a great deepening in relative sea level over a very short period of time. Relatively deep-water olenid trilobites and graptolites appear within a metre or so of the contact in the basal part of the Olenidsletta Member of the Valhallfonna Formation, in black limestones and shales, and all the typical trilobites of the Kirtonryggen Formation disappear. So striking is the change in the field that we first considered that there could be a thrust contact, but excavation has revealed a continuous section. Bathyurid trilobites do not reappear until the top of the section in the Profilbekken Member of the Valhallfonna Formation.

The graptolite fauna appearing at the base of the Olenidsletta Member of the Valhallfonna Formation immediately overlying the Nordporten Member is indicative of the Bendigonian Stage (*Tetragraptus fruticosus* Biozone) of the Australian graptolite sequence (Cooper & Fortey 1982). This correlates in turn with the early, but not earliest Floian Stage (*quondam* = early Arenigian) of the international standard Ordovician chronostratigraphy (e.g. Webby *et al.* 2004, p. 44). This graptolite evidence also proves that the time equivalents of the underlying earliest Floian, *Tetragraptus* (*Etagraptus*) *approximatus* Biozone (together with part of the lower *T. fruticosus* Biozone), should be present in the upper part of the Kirtonryggen Formation. However, there is no direct evidence for correlation between shelly and graptolitic facies that can be provided by the faunas described in this work, which are entirely comprised of species confined to shallower regions of the Ordovician Laurentian platform. The upper fauna of the Nordporten Member, the *Petigurus nero* fauna/*Lachnostoma platypyga* fauna, is surely early Floian (and Blackhillsian) in age, as deduced from its stratigraphic position beneath the *fruticosus* Biozone fauna and also as inferred from coeval faunas in western Newfoundland (e.g. Boyce *et al.* 2011). However, the location of the Tremadocian–Floian boundary within the Kirtonryggen Formation is not obvious. The *Petigurus groenlandicus* fauna underlying the *P. nero* fauna is generally very similar to the younger assemblage as far as its generic composition is concerned, although it has a distinct suite of species. It could be plausibly regarded as an earlier Blackhillsian fauna. Conodonts obtained from a sample derived from 105 m from the base of the Nordporten Member (Dr M. P. Smith, written comm. 2013) are all long-ranging Midcontinent Province species with Tulean–Blackhillsian range (*Oneotodus costatus* Ethington and Brand, *Parapanderodus striatus* (Graves and Ellison), *Tropodus comptus* (Branson and Mehl)) (e.g. Repetski, 1982) that do not refine the correlation.

However, a 2267-gm sample from the top beds of the Bassisletta Member (top of the *Chapmanopyge* fauna) provided a more indicative conodont fauna on which Dr M. P. Smith reports as follows (written comm. 2013):

'The species present are listed below (csu ranges refer to the composite standard of Sweet & Tolbert (1997)):

> *Diaphorodus* sp.
> *Drepanodus arcuatus* Pander (420–999 csu)
> *Oepikodus* sp.
> *Oistodus bransoni* Ethington and Clark 1981 (556–874 csu)
> *Oneotodus costatus* Ethington and Clark 1981 (492–906 csu)
> *Parapanderodus striatus* (Graves and Elllison) (353–1223 csu)
> *Tropodus comptus* (Branson and Mehl) (485–916 csu)
> *Ulrichodina abnormalis* (Branson and Mehl) (489–911 csu)
> *Ullrichodina simplex* (Ethington and Clark 1981) (599–864 csu)

For the most part, these are long-ranging coniform species. However, the highly distinctive *Ulrichodina simplex* first appears at a short distance above the base of the *O. communis* Zone and provides constraint on the maximum age of the sample in the composite section – in the Ibex section, this corresponds to the middle of member 3 of the Fillmore Formation (lower G2, *Protopliomerella contracta* Zone) (Ross *et al.* 1997).

Further constraint is provided by *Oepikodus* sp. that has a denticulated posterior process, a moderately inflated basal cavity and very long adenticulate anterior and lateral processes. It is similar to, but not conspecific with, *O. communis* and *Oepikodus* sp. A of Smith (1991). Age constraint is provided by the genus *Oepikodus*. At Cow Head, where conodont and graptolite ranges may be readily correlated, *Oepikodus* species appears first in the uppermost part of Bed 9, whereas the FAD of *Tetragraptus approximatus* is in the lowermost part of the 50-m-thick Bed 9 (Stouge & Bagnoli 1988). No *Oepikodus* species are known below the base of the Floian, and the sample may be referred to the Floian.'

The conclusion can be drawn that the upper Tulean fauna of the top part of the *Chapmanopyge* fauna is also close to earliest Floian, which is consistent with the type Tulean sequences in the Great Basin, western USA. The *Petigurus groenlandicus* fauna soon replaces the underlying *Chapmanopyge* fauna. Styginid, proetid and illaenid trilobites appear within the same interval, coinciding with a general increase in species-richness (Fig. 2). The appearance of more widespread conodont species and also reported by Dr Smith (also Fortey & Barnes 1977) would be consistent with this. In that the carbonate platform should respond passively to eustatic sea level changes, and it is reasonable to infer that this increase in diversity corresponded with a relative flooding event. The lithological change to include 3-m-thick units of grey, coarsely crystalline bioclastic grainstone is consistent with this interpretation. This part of the section would repay more detailed sampling than has been carried out at present. The base of the Floian (Arenig) has long been recognised as coinciding with a eustatic sea level rise (Fortey 1984; Nielsen 2004) introducing faunal change. The much more drastic change at the base of the Valhallfonna Formation is likely to be the result of the shelf foundering, because there is no event on the eustatic curve that could account for such a dramatic change in litho- and biofacies.

Correlation with other areas of Ordovician Laurentia

The succession of trilobite faunas in the Kirtonryggen Formation permits correlation with other Ordovician carbonate formations along what remains of the eastern margin of the Laurentian palaeocontinent, as seen by modern coordinates (we maintain this usage throughout this work). Previously described Early

Ordovician trilobites are often known from small samples, and their localities are scattered. However, the comparisons below are made on the basis of identical species and are reasonably secure, especially where more than one common species is known. The correlative formations are arbitrarily considered in the order of their approximate proximity to Spitsbergen. It should be noted at this point that many authors sub-divide Spitsbergen into terranes that may have been well separated in the Early Ordovician. A major transcurrent fault is posited dividing eastern from western Spitsbergen (Harland 1997 and references therein), and the eastern terrane includes the Ordovician successions with which we are concerned. Continental reconstructions of Ordovician Laurentia (e.g. Cocks & Torsvik 2011, fig. 8) favour an Ordovician position of this eastern Spitsbergen terrane adjacent to northeastern Greenland. We have no reason to question such a placement, although it will be clear from the discussions below that the trilobites are sufficiently widespread to fail to be critical in assessing palaeogeography, other than proving certainly that the Kirtonryggen Formation was deposited as part of the east (present geography) Laurentian bathyurid 'province' and sedimentary regime (Poulsen 1951; Swett 1981). Greenland terminology follows Smith & Bjerreskov (1992).

Greenland

Eastern North Greenland. – A small collection from the Wandel Valley Formation was briefly described and illustrated by Fortey (1986). The fauna includes *Petigurus nero*, *Punka flabelliformis*, *Bathyurellus abruptus* and *Benthamaspis* species all known from the *Petigurus nero* fauna from the Upper part of the Nordporten Member of the Kirtonryggen Formation and from the Catoche Formation of western Newfoundland (Fortey 1979; Boyce 1989). On the basis of Boyce's divisions of the latter, the Greenland fauna would correlate with his *Strigigenalis caudatus* Biozone, within the Blackhillsian Stage of the Ibexian in terms of the Great Basin standard.

Eastern Greenland. – The Kap Weber Formation yielded a number of trilobites to the Lauge Koch Expedition that were described by Poulsen (1937). Many are fragments, and it is not certain that they were collected from one horizon. However, in this work, we suggest that the following species are in common between the lower fauna of the Nordporten Member, *Petigurus groenlandicus* fauna, and that of the Kap Weber Formation described by

Poulsen (1937): *Petigurus groenlandicus*, *Bolbocephalus convexus*, *Bathyurellus diclementsae* n. sp. and *Licnocephala brevicauda*. Some of these determinations rely on identifying our more complete material with Poulsen's often incompletely preserved type specimens, which is always open to the possibility of error. The discussions below will provide more detail under the species concerned. However, it seems likely that the co-occurrence of several species together permits some confidence in the correlation and that Poulsen's fauna most probably originates from within a relatively short stratigraphic interval. This is the fauna below the widespread *Petigurus nero* fauna, and one that has yielded a wider range of species from Spitsbergen. *Petigurus groenlandicus* itself has a long stratigraphic range, but its co-occurrence with the other species listed is within the lower part of the Nordporten Member. Hansen & Holmer (2011) also recognised Poulsen's (1937) brachiopod species *Archaeorthis groenlandicus* in Spitsbergen. Correlation outside the eastern Laurentia is more problematic, but this fauna is regarded as equivalent to early Blackhillsian, a determination consistent with new conodont evidence.

A stratigraphically earlier fauna described by Poulsen (1937) is probably older than that from the Spora Member of the Valhallfonna Formation, *Svalbardicurus delicatus* fauna. Revision of *Hystricurus nudus* Poulsen, 1937, shows that it is a species of *Millardicurus* and likely of Skullrockian age (McCobb *et al.* in press).

Cowie & Adams (1957) collected trilobites from a fuller sequence through the Kap Weber Formation than the original collections described by C. Poulsen (1937); these are currently under study by Lucy McCobb of the National Museum of Wales. Preliminary reports suggest (McCobb *et al.* 2011) the presence of younger Blackhillsian faunas in this part of northeast Greenland, that is, faunas equivalent to the top of the Kirtonryggen Formation and the overlying Valhallfonna Formation.

Western North Greenland. – A limited trilobite fauna described from the Poulsen Cliff Formation (Fortey & Peel 1990) includes *Chapmanopyge sminue* and therefore invites comparison with the occurrence of a very similar species at the top of the Bassisletta Member and in the basal bed of the Nordporten Member in Spitsbergen. The fauna of the Poulsen Cliff Formation is therefore older than that typical of the faunas with *Petigurus* in Spitsbergen. It may equate with the younger part of the Tulean Stage of the standard Ibexian divisions.

Canada

Ellesmere Island. – Poulsen (1946) described the trilobites of the Oxford University Ellesmere Island expedition. The collections are important as it includes the type species of *Benthamaspis* and *Grinnellaspis*. Both of these genera are known from the Nordporten Member, and the type species of the former is possibly identical to the species from the *Petigurus nero* fauna. If so, it is likely that equivalents of the *Petigurus nero* fauna are present in Ellesmere Island. A recent account of the Baumann Fiord Formation trilobite fauna (Adrain & Westrop 2005) is based on more and better material, which again is a typical Bathyurid Biofacies fauna. Several species serve as a link with Spitsbergen and the Kirtonryggen Formation, especially the topmost Bassisletta Member, but the evidence is not strong. From Spitsbergen *Ceratopeltis* cf., *C. batchensis* is compared tentatively with an Ellesmere Island species, and a pygidium of *Jeffersonia viator* n. sp. is similar to one from Baumann Fiord (Adrain & Westrop 2005, figs 10.15–10.16), while *Licnocephala sanddoelaensis* Adrain & Westrop, 2005, would now be placed in *Chapmanopyge*, and apparently has related species in the Kirtonryggen Formation; these together inviting comparison of the Baumann Fjord fauna with the top of the *Chapmanopyge* fauna in Spitsbergen.

Western Newfoundland. – Platform limestones of Early Ordovician age assigned to the St George Group outcrop along much of the western coast of Newfoundland, with representative successions on the Port-au-port Peninsula, extending northwards along the coast of the Great Northern Peninsula to Cape Norman, except where interrupted by the shelfward obduction of the Cow Head Group. About halfway up the Peninsula, around the village of Port au Choix, well-exposed bedded limestones of the St George Group were visited *via* access from the sea by early collectors (including Richardson) and yielded the trilobite and mollusc specimens described by Elkanah Billings in the early years of the Canadian Geological Survey (Billings 1865). Many of these species provide the senior names for Ibexian bathyurid trilobites. In the south, the Port-au-port Peninsula has been studied particularly for the sequence of conodont faunas obtained by dissolution of the Ordovician limestones (Ji & Barnes 1994), with trilobite-based correlations added by Boyce & Knight (2010). Boyce *et al.* (2011) have shown that early Tremadocian (Skullrockian) trilobites are present in the Watts Bight Formation on the Port-au-port Peninsula, which are older than any proved from northern Spitsbergen. Further north, the trilobites first studied by Billings from the Catoche Formation in the vicinity of Port au Choix were redescribed by Fortey (1979). They include many species identical with those of the *Petigurus nero* fauna of the upper part of the Nordporten Member of the Kirtonryggen Formation in Spitsbergen, described herein. These include the following: *Petigurus nero* (Billings), *Bolbocephalus convexus* (Billings), *Bathyurellus abruptus* Billings, *Punka flabelliformis* Fortey and *Benthamaspis gibberula* (Billings), and *B. conica* Fortey. This Blackhillsian fauna is identical between these distant localities, and they must be of the same age. Boyce (1989) extended stratigraphical studies based upon trilobites into older strata in western Newfoundland and introduced a series of zonal names to summarise trilobite species distributions. He described finds from the lowest beds of the Catoche Formation and the underlying Boat Harbour Formation. In further close comparison with the Spitsbergen sequence, we here identify Boyce's *Leiostegium proprium* 'interval zone' (Boyce & Stouge 1997) in the lower member of the Boat Harbour Formation of Newfoundland with the Spora Member at the base of the Kirtonryggen Formation. The zonal fossil is related to *L. spongiosum* n. sp. in Spitsbergen; our type species of *Svalbardicurus* n. gen. is closely related to *S. seelyi* from western Newfoundland and a coarsely tuberculate hystricurid close to, if not identical with, the species named by Boyce (1989) as *Hystricurus oculilunatus,* Ross, which Boyce (1989) records from his overlying '*Randaynia saundersi* interval zone'.

Between these two correlatable faunas lies a stratigraphic interval in which there is less comparison between the Early Ordovician of Spitsbergen and western Newfoundland. In the Spitsbergen sequence, this interval includes the Bassisletta Member and the lower part of the Nordporten Member, a thickness of more than 300 m of carbonates. Boyce (1989, fig. 7) marks a hiatus (marked by pebble beds) in the Newfoundland succession to include at least some of this interval. In Spitsbergen, the Bassisletta Member comprises a considerable thickness of shallow-water dolomites that may equate with a period of non-deposition in Newfoundland. Our *Peltabellia glabra* fauna fits within this 'gap', and there is no direct evidence for its time equivalent in western Newfoundland. Nor does Boyce (1989) record trilobites suggestive of the *Chapmanopyge* fauna. So it is likely that there is a period of non-deposition and/or erosion, or at least unfossiliferous strata, which renders the western Newfoundland platform succession incomplete. Following the Newfoundland hiatus, Boyce's (1989) *Strigigenalis brevicaudata* 'lineage zone' underlies an undoubted equivalent of the *Petigurus nero* fauna in Spitsbergen, which is assigned to Boyce's *Strigigenalis caudata* Zone. This underlying interval should, therefore, include the equivalents of the *Petigurus groenlandicus* fauna in the Nordporten Member in Spitsbergen. We do not have the Newfoundland zonal fossil (*S. brevicaudata*) in Spitsbergen, and the only species in common is *Uromystrum affine* (Poulsen) (see p. 83 for discussion) and *Ischyrotoma parallela* Boyce; generally, a more diverse fauna is present in Spitsbergen than in Newfoundland through this interval, one that includes the earliest *Illaenus*, styginid and proetid records. Boyce & Stouge (1997) matched conodont and trilobite occurrences through the same Newfoundland successions and concluded that the boundary between Floian (= basal Arenigian) and Tremadocian strata was at the base of their *Strigigenalis brevicaudata* 'lineage zone'. This is compatible with our view that the Tremadocian–Floian boundary lies somewhat below the base of the *Petigurus groenlandicus* fauna in Spitsbergen.

In summary, there are two good stratigraphical ties near the top and bottom of the Kirtonryggen Formation into the succession of western Newfoundland. The mid-part of the Kirtonryggen Formation includes faunas as yet unrecognised in Newfoundland, indicative of a stratigraphic gap there ('Boat Harbour unconformity').

It should be recalled here that the Early Ordovician stage names used by Boyce (1989) and Boyce & Stouge (1997) in Newfoundland were divisions of the classical 'Canadian Series' – in ascending order, Gasconadian, Demingian, Jeffersonian and Cassinian – based originally upon rock successions from various parts of eastern Laurentia. Nautiloid faunas (e.g. Flower 1964, 1968) were crucial to the understanding of these divisions, and the virtual disappearance of students of this important group of molluscs, as well as a wish for uniformity, has contributed to a comparative neglect of this scheme of local chronostratigraphy. Modern usage attempts to correlate with the standard Ibexian stage nomenclature based upon the successions in the Great Basin, western USA (Ross *et al.* 1997), despite the differences in the faunas there from those on the eastern side of the Laurentian continent. Taylor *et al.* (*in* Derby *et al.* 2012, p. 22) have advocated continued use of the older-stage terms for at least the upper part of the Ibexian.

Mingan Islands, Quebec. – A brief account of trilobites in the Mingan Islands by Twenhofel (1938) has been recently updated by Shaw & Bolton (2011). The Romaine Formation includes a meagre fauna that can be compared with one from the earlier part of the Kirtonryggen Formation. We have cautiously identified a species originally described by Billings (1859) as *Bathyurus amplimarginiatus* as present in the Bassisletta Member and referred it to *Chapmanopyge*. Shaw & Bolton (2011) assigned this species to *Peltabellia*. It is accompanied by a *Bolbocephalus* species. If these determinations are correct, it indicates the presence of Tulean strata and points to the absence in the Mingan Islands of the *Petigurus nero* fauna, which is otherwise widespread. The overlying Mingan Ordovician trilobite faunas all prove much younger Ordovician ages than the Kirtonryggen Formation.

Northwest Scotland

Fortey (1992) summarised what little is known about the trilobites from the thick shallow-water limestone succession that extends in a narrow band of outcrop running from the Isle of Skye in the south to Durness in the north. The presence of *Petigurus nero* is evidence for the equivalents of the younger fauna of the Nordporten Member, Kirtonryggen Formation, in Scotland. A free cheek of a *Jeffersonia* species is like that illustrated here in Figure 10C from lower in the same member. This is suggestive of a spread of age equivalents to the Spitsbergen succession in northwest Scotland, although the evidence from trilobites is sparse. However, a succession of nautiloid faunas (Evans 2011) proves the existence in Scotland of a relatively complete Ibexian succession.

USA

New York State and Vermont. – This is the type area of the 'Cassinian Stage' of the Canadian Series, which has now been superseded terminologically by the Blackhillsian Stage of the Ibexian (also = early Floian and = early Arenigian). Brett & Westrop (1996) revised the relevant trilobite fauna of the Fort Cassin Formation and proved its undoubted correlation with the Catoche Formation of western Newfoundland. It equates in turn with the upper part of the *Petigurus nero* fauna in Spitsbergen, although the only species in common is probably one of *Benthamaspis* (see p. 62). The underlying Rochdale ('Fort Ann') Formation includes *Svalbardicurus seelyi*, which is closely related to the type species of that genus, described below, from the Spora Member of the Kirtonryggen Formation. Boyce (1989) recorded *S. seelyi* from the lower member of the Boat Harbour Formation in western Newfoundland. These occurrences are likely of similar age, depending on the eventual stratigraphic range of species of *Svalbardicurus*. The equivalents of the *Petigurus groenlandicus* and *Chapmanopyge* faunas have not been proved in this area of the eastern USA. Recently, Landing *et al.* (2012) have described additional trilobites from the Rochdale Formation, which includes middle–late Stairsian trilobites that invite comparison with those of the Spora Member; these authors correlate the appearance of this fauna with a Tremadocian eustatic sea level rise. Their 'Hystricurus' sp. nov. might also be better accommodated within *Svalbardicurus* n. gen., while a coarsely tuberculate hystricurid is also present alongside a species of *Leiostegium* similar to that from Spitsbergen. It seems plausible that the same eustatic event is reflected in a similar fauna.

Oklahoma and Missouri. – Loch's (2007) account of the trilobite fauna of the Kindblade Formation of Oklahoma permits correlation with the upper part of the Kirtonryggen Formation, although there are relatively few species in common, and several taxonomic questions emerge. At the top of the succession in Oklahoma, the *Strigigenalis caudata* Zone is stated to be the equivalent of the interval carrying the same name in western Newfoundland, where many elements of our *Petigurus nero* fauna are also known to be present. Faunas underlying the *S. caudata* Zone in Oklahoma should, therefore, include equivalents of the *Petigurus groenlandicus* fauna. However, Loch's (2007) *Bolbocephalus stitti* Zone does include a few species in common with its presumed equivalent in Spitsbergen. We have identified our new species *Bolbocephalus gunnari* with Loch's *Bolbocephalus* sp. 2 (see p. 31), which Loch (2007, p. 28) states has a range extending through the zones underlying the *B. stitti* Zone: that is, *Benthamaspis rhochmotis* Zone extending into the *Petigurus cullisoni* Zone. However, in his summaries of zonal species distributions, it is listed (Loch 2007, p. 11) from the *P. stitti* Zone, but from neither of the underlying two zones. The species has a long range in the lower part of the Nordporten Member. A second *Bolbocephalus* species, here recorded as *B. stclairi* Cullison, is apparently identical to a cranidium figured by Loch (2007 pl. 7, fig. 8) also from the *B. stitti* Zone (Loch 2007, p. 85). More problematic is the species Loch (2007) named as *Gelasinocephalus pustulosus* (see also pp. 54, 89). We regard this species as a dimeropygid closely related, if not identical, to *Ischyrotoma parallela* Boyce, 1989, which we record from the lower part of the Nordporten Member herein. Given the ambiguities involved with these three species, it is not possible to be more precise than to say that some part of the *B. stitti* Zone, and/or *Petigurus cullisoni* Zone and *B. rhochmotis* Zone in the Kindblade Formation, Oklahoma, includes equivalents of the *Petigurus groenlandicus* fauna that

extends through the lower part of the Nordporten Member in Spitsbergen.

As in the Bassisletta Member in Spitsbergen, in Oklahoma, there is an underlying interval in which the common bathyurelline is *Chapmanopyge* (see p. 63). Loch (2007) divided this interval into two zones: *Rananasus brevicephalus* Zone (below) and *Jeffersonia granosa* Zone (above). The top of the Bassisletta Member has yielded a species of *Chapmanopyge*, which we consider the same as Loch's *Chapmania* cf. sp. 1, a species from the *J. granosa* Zone. A small collection made 140 m below that yielding *Chapmania* cf. sp. 1 has sparse material attributed to *Chapmanopyge* 'amplimarginata' (Billings), a species closely similar to *C. taylori* Loch, 2007, from his *R. brevicephalus* Zone. It seems likely that the upper half of the Bassisletta Member may equate to at least parts of the earliest two zones recognised by Loch (2007) in the Kindblade Formation. This includes our *Chapmanopyge* fauna. With the exception of the uppermost beds, this fauna is poorly represented by species in Spitsbergen, and trilobites have not been recovered from much of the very shallow-water, dolomitic succession. This interval of the Ibexian is much richer in species in Oklahoma, where earlier representatives of genera that were to become significant in the Nordporten Member have been recovered.

Further correlation into the Ozark Uplift, Missouri, still depends on the work of Cullison (1944), who illustrated many trilobites (mostly internal moulds) at rather small size. A few originals of his *Jeffersonia* species are illustrated herein. This is the type area of the previously used Jeffersonian Stage of the Canadian Series. Loch (2007) reillustrated a number of the species in common with the Kindblade Formation on which correlation largely depends, and his summary (Loch 2007, text-fig. 7) cannot be refined by any of our discoveries. Broadly speaking, the Cotter Formation of Cullison (1944) includes the equivalent of the *Petigurus nero* fauna (see also Boyce 1989) and the upper part of the Kirtonryggen Formation. Thompson (1991) has changed the stratigraphical nomenclature in Missouri, such that this interval is now termed the Bull Shoals Member of an expanded Cotter Formation. Two further members below that (formerly Theodosia Formation of E. O. Ulrich and Cullison) presumably include the equivalents of the lower part of the Kirtonryggen Formation and the *P. groenlandicus* fauna. Faunas with *Chapmanopyge* are found in the still older Rich Fountain Formation of Cullison (1944) (= Jefferson City Formation of Thompson, 1991). Our only addition to correlative data is the identification of *Jeffersonia viator* n.

sp. herein with *Jeffersonia* sp. 1 of Cullison (1944) (top of Bassisletta Member and low 'Theodosia Formation', respectively) and *Bolbocephalus stclairi* Cullison, 1944 ('Cotter' and Nordporten Member).

Occurrence of trilobites in the Kirtonryggen Formation

Field occurrence

The trilobites of the Kirtonryggen Formation are invariably disarticulated sclerites. These fragments have clearly been reworked and concentrated into shell lag deposits or storm accumulations that may be discovered in the field as small lenses, with irregularly orientated fragments. This makes for special difficulties, in that robust pygidia and free cheeks tend to be preserved more frequently than cranidia, which are vulnerable to breakage along the posterior limbs of the fixigenae. In some cases, we have not found the appropriate cranidia associated with very distinctive pygidia, which may be the consequence of sorting. In a few beds, such as that at 105 m from the base of the Nordporten Member, there is a thicker, fossiliferous horizon from which it is possible to extract more information on individual species, including hypostomes. Where blocks of such a horizon have been collected, much of the work has been done by breaking it up into pieces in the laboratory to recover sclerites that were inconspicuous in the field.

In spite of their fragmentary nature, beautiful surface sculpture is usually preserved on the trilobite fragments. This implies that they were not eroded before burial and suggests rather rapid concentration by tides or storms into pockets on the carbonate platform not far from where the animals originally lived. Not all the bathyurids had thick cuticle (e.g. *Bathyurellus*, *Harlandaspis*), but others were noticeably robust, such as *Jeffersonia* and *Petigurus*. Many species had more or less tuberculate glabellas, which may have contributed to cuticle strength in life. This also means that it is difficult to determine which species is present from sight of the glabella alone. Extraction of good material can be hindered by patchy dolomitisation that has the effect of obliterating trilobite cuticle. Some horizons have cherty grains that have a similar result in halting extraction. Taken together, such preservation details mean that mechanical preparation of the trilobites illustrated in this work has taken several years.

The clear evidence of sorting obviously imposes limitations on summaries of alpha diversity. However, since the inshore biofacies represented in the Kirtonryggen Formation are localised on the carbonate platform, it is not unreasonable to suppose that the commonest species in the sorted sclerites were also the most abundant in the original fauna. Most of the collections were far too selective in the field for 'good finds' to be representative, but three of the bulk rock collections broken up in the laboratory allow for counting relatively large numbers of sclerites. These are as follows: the oldest Ordovician fauna of the Spora Member; that of the mid-Bassisletta Member; and the *Petigurus groenlandicus* fauna from the lower part of the Nordporten Member. In practice, the Bassisletta Member fauna consists entirely of the single species *Peltabellia glabra*.

The Spora Member fauna shows the dominance of the hystricurid *Svalbardicurus*, with *Leiostegium* and *Eurymphysurina* also numerous. Identifiable sclerites (cranidia, pygidia) are as follows: *Svalbardicurus delicatus* 29; *Leiostegium spongiosium* 20; *Eurysymphysurina spora* 17; *Hystricurus* sp. B 4; *Hystricurus* sp. 1 4; and *Pilekia* cf *P. trio* Hintze 2. The Nordporten Member fauna at 105 m from the base has the bathyurids *Uromystrum* and *Jeffersonia* predominant, with the curious species *Harlandaspis elegans* common. The recognisable sclerites group as follows: *Uromystrum affine* 75; *Jeffersonia striagena* 37; *Harlandaspis erugata* 29; *Ischyrotoma parallela* 10; *Psalikilopsis* n. sp. 7; *Ischyrotoma parallela* 7; *Punka latissima* 3; *Petigurus groenlandicus* 3; *Licnocephala* spp. 3; *Carolinites* sp A. 3; others 5. *P. groenlandicus* is a very large species, which is why it is readily spotted in the field, but it is relatively rare in 'crack out'. It is also interesting to observe a marked convergence between cephalic morphology of the unrelated trilobites *Uromystrum* and *Eurysymphysurina*, which are both illaenimorphs in the terminology of Fortey & Owens (1990). Both *Leiostegium* and *Harlandaspis* have long glabellas and reduced pre-glabellar fields. *Svalbardicurus* and *Jeffersonia* have more generalised morphologies.

Biofacies and biogeography

The fauna of the Kirtonryggen Formation is dominated by the inshore carbonate Bathyurid Biofacies, typified by a profusion of the eponymous family, which was confined to palaeotropical regions in the Early Ordovician (e.g. Whittington 1963; Fortey & Cocks 2003). Evidence of the shallow-water nature of the palaeoenvironment is provided by the ubiquity of edgewise conglomerates, stromatolites and limestones with fossil algae, and abundant porcellanitic dolostones with occasional oolitic beds in the Bassisletta Member. Almost all the trilobites are disarticulated and/or reworked. In the lower part of the Bassisletta Member, *Peltabellia* is associated with stromatolite heads, and it is likely that this trilobite occupied channels between these structures. Otherwise, this part of the succession is lacking trilobites, and it is probable that it was too shallow and/or too saline for them to thrive. As outlined above, the Spitsbergen succession is the northernmost of many occurrences of the Bathyurid Biofacies along the eastern side (present geography) of Laurentia, in a belt stretching from New Mexico, the Ozark Uplift, New York State, Ontario and Quebec, western Newfoundland, northwest Scotland and East and North Greenland. It is everywhere associated with an extensive carbonate platform (Derby *et al.* 2012). This distribution is confirmed by the detailed faunal comparisons listed in the systematic part, where appropriate references are cited. The endemic trilobites are associated with a typical Mid-continent conodont fauna, together with characteristic nautiloids, gastropods and sponges. There is evidence of a similar fauna in platform northeast Siberia, where *Biolgina* and other bathyurids occur.

In the earliest part of the Ordovician succession, the earlier Ibexian (Stairsian) Spora Member of the Kirtonryggen Formation pre-dates the radiation of the Bathyuridae. However, a shallow-water biofacies that is typified by hystricurids, leiostegiids and Symphysurinidae, and very similar to one of the same age described from western Newfoundland by Boyce (1989), seems to be no different in lithological characters from strata that would be typical of Bathyurid biofacies higher in the succession. There appears to have been a fundamental change in the composition of trilobite faunas between the Stairsian and the Tulean + Blackhillsian. Ethington *et al.* (1987) noticed a comparable change in the composition of conodont faunas of the Mid-continent Province.

At the upper limit of the succession, the change at the base of the Valhallfonna Formation into the Olenid biofacies is profound, being marked by the replacement of the shallow-water limestones of the Kirtonryggen Formation by black shales and well-bedded limestones bearing abundant graptolites alongside the characteristic deeper-water olenid trilobites. The transition is quite abrupt. The top one metre of the Kirtonryggen Formation differs faunally from the beds below in having numerous large asaphids. Possibly, this is the consequence of the beginning of a deepening event. There are also numerous current-oriented orthocone nautiloids in the same interval, which Brett (1995) notes commonly occur in condensed intervals associated with flooding surfaces. Graptolite stipes appear for the first time. It seems unlikely that the drastic change from Bathyurid Biofacies to Olenid Biofacies across

the formational boundaries is accountable by a eustatic event alone. Sea level curves for the Ordovician (e.g. Nielsen 2004) do not record a particularly major event within this level of the Ibexian (early Floian/Arenigian). It seems more likely that the shelf foundered in this part of the northern Laurentian carbonate platform, allowing the sudden appearance of hitherto exterior biofacies in the rock record. The rich Olenid Biofacies of the Valhallfonna Formation is unique at the time of writing – it is not known elsewhere along the eastern margin of Laurentia, possibly because it is overridden in the parautochthonous belt of the Appalachian chain. In Spitsbergen, it may have been preserved because of a unique event affecting the shelf edge there.

Peculiarities of the Spitsbergen faunas

Although the fauna of the Kirtonryggen Formation is developed in inshore, low-latitude carbonate biofacies, there are some interesting differences from other contemporary trilobite faunas that apparently lived in similar palaeoenvironments. There are some genera and even families that we might expect to be present that are not known in Spitsbergen, in spite of our splitting up large bulk rock collections specifically to search for elusive trilobites. Most remarkable among these cases is the complete absence of trilobites of the Family Pliomeridae. These trilobites are very diverse in the Great Basin, as a series of recent papers by Dr J. M. Adrain and his colleagues have shown. However, pliomerids do extend – albeit in lesser variety – into the eastern section of Laurentia, from mainland Canada through Newfoundland as far as northern Greenland (Fortey & Peel 1990). They do eventually appear in the upper Profilbekken Member of the Valhallfonna Formation in the Middle Ordovician, but the Lower Ordovician of Spitsbergen is without them. As noted, the fauna of the Kirtonryggen Formation can be often matched species for species with the St George Group in western Newfoundland. However, one bathyurid genus, *Strigigenalis*, which is relatively common in Newfoundland southwards to Oklahoma, USA, has not been found in Spitsbergen, in spite of the abundance of other bathyurid species. Boyce regarded species of this particular genus as of special stratigraphic significance and employed them as biozone name bearers in Newfoundland. *Strigigenalis* species develop a striking posterior pygidial spine. In the Great Basin, it has a homoeomorph in the genus *Gladiatoria* (Adrain *et al.* 2011a). It is noted that *none* of the species we have recovered from the Kirtonryggen Formation have pygidial spines extending backwards. In other contemporary Laurentian faunas, such spines are developed in several other bathyurid genera, as well as in asaphids. The latter family is numerous in Great Basin localities, and asaphids also appear at the very top of the Kirtonryggen Formation associated with an obvious facies change: a deepening as the Valhallfonna Formation is approached. They were presumably adapted to more open shelf conditions at the edge of the carbonate platform.

The absence of pliomerids and *Strigigenalis* through the considerable thickness of the Valhallfonna Formation is harder to explain. Presumably some environmental control was responsible, but it must have been a subtle one. As noted, indicators of very shallow or intertidal conditions such as stromatolites and porcellanitic dolostones, are numerous in the Spitsbergen succession. It does seem possible that shallow, tropical platform conditions promoted elevated salinities and/or temperatures to which bathyurids were particularly well adapted. Sclerites of the large-sized genus *Petigurus*, with very thick cuticle, can usually be found even in the absence of all other trilobites, and this genus may have had the greatest tolerance range of all. *Bathyurus* appears to have been similarly adapted later in the Ordovician. Gastropods, now largely recrystallised and hard to extract from the rock, are frequently associated with *Petigurus* fragments and may well have been algal grazers spending active periods on intertidal flats. A possible scenario might suggest that Pliomeridae and *Strigigenalis* avoided such specialist conditions. Certainly, where pliomerids are found in Ordovician strata elsewhere in the world – as in Baltica or Avalonia – they are associated with open shelf deposits and normal marine conditions.

Eastern and western Laurentia

Although the early Ordovician trilobite fauna of the Kirtonryggen Formation is generally typical of the palaeoequatorial Bathyurid 'Province', there is evidence of differentiation of faunas between the eastern and western (current geography) sides of the Laurentian palaeocontinent, as Loch (2007) has noted. The succession in the Great Basin on the western side provides the stratigraphical standard for sub-divisions of the Ibexian 'Series' (Ross *et al.* 1997), a designation that has replaced the old term 'Canadian', which was based upon a scattered collection of type sections on the current *eastern* side of Laurentia. The term 'Tremadocian' has been adopted as the preferred name for the earliest chronostratigraphic sub-division of the Ordovician by the Ordovician Subcommission, but 'Ibexian' (and

its constituent stage divisions) is still appropriate for use as the regional standard for the North American continent, since faunal connections with the European Tremadocian are minimal. Ibexian faunas of the Great Basin are known from the classic works of Ross (1951) and Hintze (1953). Subsequent and important refinements have been introduced, for example, by Adrain *et al.* (2009), the latter paper having introduced a sequence of finely drawn biozones to replace some of the 'letter zones' (Zone A, Zone B, etc.) of Ross (1951), and their successors named after index-species (Hintze 1973). The extension northwestwards of these faunas into the Rocky Mountains of Alberta, Canada, has been documented by Dean (1989), where a very similar succession of trilobites occurs.

However, none of the name-bearing trilobites typifying the Ibexian Great Basin – western Canadian biostratigraphic scheme is known from eastern Laurentia. Even allowing for the peculiarities of the northern Spitsbergen faunas listed above, the trilobite faunas described from Utah, Nevada and Idaho are different from those from the Ordovician outcrops running from Oklahoma northwards through New York, eastern Canada, Greenland and Svalbard discussed above. While the common presence of bathyurid trilobites, the old Bathyurid 'Province' (Whittington 1963), provides a linking factor between eastern and western Laurentia, there are noticeable differences at generic and species level. For example, the conspicuous, large Blackhillsian bathyurid *Petigurus* is found in many eastern Laurentian sections, and one species, *P. nero*, is described from northwest Scotland, western Newfoundland, Greenland and Svalbard, where it is one of the most conspicuous trilobites to be recognised in the field. It does not seem possible that collection failure alone could be responsible for its apparent absence in strata of comparable age in the western USA. It does, however, appear in this region at the base of the Middle Ordovician, Whiterockian (Fortey & Droser 1996), which was a time of general faunal turnover. Similarly, species of *Uromystrum*, *Harlandaspis*, *Chapmanopyge* and *Punka* are predominantly eastern Laurentian in the Ibexian, and as shown in the systematic part, they form clades of closely related species. The asaphid *Isoteloides* is the usual representative of its family from eastern Laurentia, but in western Laurentia, six or more genera of the same family are known and often abundant. We should perhaps remark here that *Isoteloides* has not turned up in the Kirtonryggen Formation, so it is not ubiquitous even in eastern Laurentia. On the contrary, many Tulean–Blackhillsian genera from the Great Basin currently being redescribed by

Dr J. M. Adrain and his colleagues, such as *Psalikilus*, *Gladiatoria*, *Pseudoolenoides* and *Aponileus*, are rare or yet to be recognised along the eastern belt. These differences ought to be reduced by further collecting to secure rarities in either region, but they are striking in the context of a single palaeocontinent. A number of genera of more general distribution offer possibilities for correlation based on trilobites. *Benthamaspis* is widely reported, as are the dimeropygids *Ischyrotoma* and *Dimeropygiella*, and it seems likely that the early representatives of the pelagic telephinids such as *Carolinites* described below will provide the necessary species-level links for precise correlation. At the moment, however, there will remain uncertainties in the application of the fine zonation currently being worked out in the Great Basin to eastern platform successions. Conodonts of the Mid-continent Province are more widespread and provide useful temporal links between the two sides of Laurentia (e.g. Ji & Barnes 1994; Ji & Barnes 1996), although it is likely that the trilobites will ultimately provide the finer calibration.

The possible reasons for the macrofaunal differentiation between the two sides of Early Ordovician Laurentia are unproven. On reconstructions of Ordovician Laurentia at this time (e.g. Cocks & Torsvik 2011, for a recent version), the palaeoequator runs along the spine of the palaeocontinent, so what is referred to as 'eastern' and 'western' Laurentia above should really be 'southern' and 'northern' on the original geography. The Mid-continent provides a barrier between the two sides, even though similarly low latitudes are present on the continental margins to either side of the palaeoequator. The question is whether this divide was sufficient to allow trilobites on each side to follow separate evolutionary trajectories from shared ancestors (Bathyuridae are considered to be monophyletic) or whether different environments were established on the two seaboards that resulted in local speciation and specialisation. The fact that some benthic genera, such as *Benthamaspis*, *Bolbocephalus* and *Ischyrotoma*, are found on both sides of the divide does carry the implication that any barrier to dispersal was 'porous' and favours some environmental factors as an explanation. The same argument has been made above concerning the rarity of Pliomeridae. The sedimentology and palaeoenvironments of the (current) eastern Laurentian margin have been investigated for thirty years, and the properties of the Early Ordovician carbonate platform are well-established. There are local variations, but such features as thrombolitic 'bioherms' and sponge reefs, stromatolites and algal laminates are widespread, for example, as described from the St George Group, western Newfoundland,

by Pratt & James (1986, 1989) with similar lithologies spread along the Appalachian chain southwards at least as far as Oklahoma (Riding & Toomey 1972; Toomey & Nitecki 1979). The trilobites and sedimentary features of the St George Group in Newfoundland, and the Kirtonryggen Formation in Spitsbergen, are closely comparable, and it is reasonable to suppose that similar ecological niches were developed in the latter. This patchwork of intertidal and shallow sub-tidal habitats may have been different from the open shelf environments developed in the Great Basin and elsewhere in the western USA and Canada. The same inshore habitat on the east supported great numbers of gastropods (e.g. western Newfoundland, Rohr *et al.* 2001) including large macluritids and the operculum *Ceratopea*, with individual species such as *C. unguis* being widespread over the vast platform (Yochelson & Bridge 1957; Yochelson & Copeland 1974). These molluscs included grazers on microbial mats and algae. While it is possible that the specialised trilobites that lived nearby adopted similar habits if their basipodal spines could be modified as 'scrapers', it is perhaps more likely that they pursued small prey items in intertidal channels since the morphology of the extreme inshore genera (*Petigurus*) is the phacomorph design generally associated with predatory/scavenging habits.

Origin of major trilobite clades

It is of interest that we report in this work the earliest occurrence of an illaenoid, and the earliest proetoid, and the first-known Laurentian styginoid from the inshore carbonates of the Kirtonryggen Formation. The suggestion that major marine clades may originate in inshore settings as comparatively rare elements in the total fauna is several decades old (Jablonski & Bottjer 1990; Sepkoski 1991). From an origin in shallow, disturbed habitats, it was claimed such clades spread and diversify across shelf and into slope settings progressively, where they may persist after their inshore relatives have died out. Jablonski (2005) has updated information for the origin of Mesozoic marine clades, showing that the patterns are robust in the context of changing phylogenetic methods and new discoveries. For the Palaeozoic, there is perhaps less data, but the confirmation of the pattern requires listing more individual instances and is an incentive for conducting fieldwork in rocks deriving from more sparingly fossiliferous palaeoenvironments. Illaenids and styginids have a long and successful subsequent history extending to the Devonian, in contrast to the bathyurids that greatly outnumber them in Spitsbergen succession, which did not survive the Ordovician. In both cases, we have acquired most specimens from single horizons (the *Illaenus* is rather numerous where it occurs), and they do not reappear regularly, or even again, in strata above in the same formation. The implication might be that they had very particular habitat requirements at this early stage in their history, while the bathyurids and dimeropygids occupied more, or at least more widespread ecological niches. The earliest proetoid is at the origin of a clade that outlasted all other trilobites into the Permian. Its occurrence in the Kirtonryggen Formation is more sporadic through an appreciable thickness of strata. The first appearance of all these taxa is at what we have suggested above is close to, or at the Tremadocian–Floian boundary.

The next youngest proetoid is probably that described as *Phaseolops ceryx* from the Floian Tourmakeady Limestone of western Ireland (Adrain & Fortey 1997), which is still in a shallow-water limestone of illaenid–cheirurid biofacies on the margins of Laurentia. By Middle Ordovician times, species of Proetidae are widely known from open shelf deposits flanking Laurentia, Baltica and Gondwana palaeocontinents. *Illaenus* itself is known from Dapingian strata in Baltica. By earliest Middle Ordovician times, the family Illaenidae is also well-established in comparatively deep-water strata around Gondwana and Perunica (*Ectillaenus*). Styginids are present in open shelf strata in Scandinavia (Bornholm) by somewhat later in the Floian (Poulsen 1965). Hence, if the Spitsbergen occurrences really are the first of their respective groups, colonisation of other habitats was both fast and geographically spread.

It might be added that the cheiruroid *Pilekia* (known from the Spora Member) is also the earliest member of the Cheiruroidea, although the earliest examples in the genus are probably from the Great Basin, western USA. If the taxonomic compass is extended to include other arthropods, then the Kirtonryggen Formation also includes the earliest leperditicope 'ostracod' (Williams & Siveter 2008). Considering that the fauna as a whole is not especially diverse, specialised as it was for the special conditions pertaining on a shallow carbonate platform, the first occurrence of this number of novelties is of interest. There may have been numerous opportunities for specialists in and around the thrombolitic mounds and *Renalcis* and *Epiphyton* 'reefs' (Riding & Toomey 1972) that were commonly developed on the eastern Laurentian margins. Possibly this complex of habitats favoured the circumstances that produced innovations, under such

mechanisms as outlined by Jablonski (2005). It is interesting that these (taxonomic) innovations are accompanied by a series of morphological innovations among the contemporary bathyurid trilobites, as they morphed into a bewildering set of advanced characters. It is difficult to understand why the bathyurids were short-lived genera that never moved from their specialised habitats, while their relatively inconspicuous contemporaries diversified rapidly into many habitats and outlasted the majority of other trilobites.

As a qualification to the preceding argument, it should be added that the horizons where these early representatives of major clades appear are *also* those probably corresponding with transgressive system tracts in the section. This could imply that the appearance of trilobite 'innovations' at these levels might be explained by the sudden appearance of suitable substrates/conditions suitable for taxa that had evolved elsewhere. This possibility can only be proven by the discovery of still earlier representatives of a given clade. It is of course possible that transgressive phases served equally to stimulate evolutionary novelty.

The family Bathyuridae

The trilobite family Bathyuridae includes the majority of species described in the systematic section. The taxonomy of the family is still not established on a modern footing. However, the species described in this work add many details to the morphology that will prove of service in future analyses. The Ordovician evolutionary radiation of the family on the Laurentian platform is remarkable, so much so that framing a diagnosis for the family that effectively excludes all other trilobites is hardly possible. The range of morphologies that rapidly evolved in Bathyuridae includes species that are convergent on other families. In practice, the assignment to this particular family relies on the discovery of intermediate morphologies that link a species under examination with a more typical member of the group. A great number of bathyurids display a distinctive, relatively narrow and elongate glabella, which is often cigar shaped or anteriorly pointed. Other Ordovician families can readily be separated on this feature. However, *Benthamaspis* has a squat, rectangular glabella. Bathyurids can exhibit any variation in convexity and effacement. They may have stout genal spines, or long slender ones, or lose them altogether. Pygidia vary from small to large, convex to flat, spinose or not. Surface sculpture can be of any type known in the Trilobita – or none.

Sub-division of the family into sub-groups (sub-families) should be possible, but for the moment, we shall recognise only sub-family Bathyurellinae comprising those bathyurids with relatively large, flattened fan- to paddle-shaped pygidia and often narrowly triangular, falcate free cheeks with genal spines that broadly continue in the same plane from the adjacent genal fields. They range from Ibexian to Whiterockian. However, the systematics eventually work out, this sub-family is likely to resolve as a clade. The remaining bathyurid genera are placed in Bathyurinae, a heterogeneous group ranging through most of the Ordovician, and which will very likely resolve into several clades. The primitive bathyurid morphology is represented by the genus *Peltabellia*, which is probably paraphyletic.

The problem of type species

A particular problem with bathyurid genera is the fact that their type species were often proposed in the past on the basis of inadequate material, by modern standards. Mostly, this is a matter of the type specimen being a cranidium or pygidium, with none of the other sclerites associated. Since we now know that all parts of the bathyurid exoskeleton display important taxonomic information, this practice was unfortunate, to say the least. But it has proved to be the case that subsequent research has 'rescued' a number of these genera from obscurity. According to the *Rules of Zoological Nomenclature*, all the generic names in question are validly proposed. Ideally, more material from type localities would make good the deficiencies of the original descriptions, but this has not always proved possible, not least because some of the type localities are remote and inaccessible (if, indeed, they are known precisely). A list of genera that are, or have been, inadequately characterised summarises some of the problems with the taxonomy of bathyurid trilobites, in alphabetical order, and excluding subsequent junior synonyms:

Aponileus Hu, 1963. – Based on poorly preserved material that was not even originally assigned to Bathyuridae, this genus has now been well characterised by silicified material described by Adrain *et al.* (2012).

Bathyurina Poulsen, 1937. – Based upon a single cranidium (*B. megalops*) from eastern Greenland, this large-eyed bathyurine had no free cheek or pygidium associated with it. Fortey (1979) associated a cranidium resembling that of the type species with the pygidium of *Bathyurus timon* Billings,

1865, from the Catoche Formation of western Newfoundland. However, the pygidium of *timon* type was also very similar to the type species (based on a pygidium) of *Jeffersonia*, *J. exterminata* Poulsen, 1927. Fortey (1986) subsequently reverted to using the senior name *Jeffersonia* for *Bathyurus timon*. In spite of identifying a number of Poulsen's (1937) Kap Weber Formation species from the Kirtonryggen Formation, there is no exact match for *B. megalops* in the present study, and the nomenclatural problems cannot be resolved here. The answer is to be found by making new collections of the type species from eastern Greenland, which is not forthcoming.

Ceratopeltis Poulsen, 1948. – The type species was based on a pygidium from Greenland that had not been completely prepared. It was placed among the *incertae sedis* in the 1959 Trilobite *Treatise* (Moore 1959), but Fortey & Peel (1983) recognised it as a bathyurid and described additional sclerites, and Adrain & Westrop (2005) added a further well-preserved species to the genus. Hence, it has been rescued from obscurity.

Gelasinocephalus Loch, 2007. – has type species, *G. whittingtoni* Loch, 2007, from Oklahoma that seems clearly related to *Bolbocephalus,* differing mainly in having a relatively long pre-glabellar field and a more triangular pygidium with narrower (tr.) pleural fields. The attributed free cheek (Loch 2007, plate 9, figs 12, 13) is identical to that of some species of *Bolbocephalus*. However, a second species placed in this genus by Loch, *G. pustulosus* Loch, 2007, has a tuberculate glabella, which compares with that of such dimeropygids as *Ischyrotoma*, and a free cheek, which would also be compatible with that family, and is completely unlike that of the type species. The two species only share the pre-glabellar field and an effaced cranidial border, neither of which are very diagnostic. If *Gelasinocephalus* becomes monotypic, its distinction from *Bolbocephalus* is also more problematic.

Gignopeltis Raymond, 1913. – Based on a pygidium from the 'Beekmantown', the type specimen was reillustrated by Ludvigsen (1979). This specimen could represent a senior name for either *Benthamaspis* Poulsen or *Harlandaspis* n. gen., both of which have more or less identical pygidia. Or it could be neither of them, depending on the cephalic parts (unknown). The genus has fallen into general disuse, although revived by Boyce as a zonal fossil in the St George Group. It cannot be used without identification of cephalic sclerites from the type locality.

Gladiatoria Hupé, 1955. – Based upon a silicified species described by Ross (1951) as *Macropyge* sp. and has now been well characterised as a bathyurid clade by Adrain *et al.* (2011a).

Gonioteloides Kobayashi, 1955. – Type species, *G. monoceros*, Kobayashi 1955, McKay Group British Columbia, is based on a pygidium, and nothing more is known of it.

Grinnellaspis Poulsen, 1946. – Based on one pygidium from Ellesmere Island. It has been used to include some bathyurellines with strongly radially furrowed pygidial borders, but its relationship to *Punka* is unclear, and its usage is bound to be provisional until the type species is known more completely.

Hadrohybus Raymond, 1925. – Based upon crania from allochthonous boulders in the Cow Head Group, western Newfoundland. Placed in the Family Cheiruridae in the *Treatise on Invertebrate Paleontology* (Moore 1959), Fortey (1988) showed that it was, in fact, a bathyurid allied to *Bolbocephalus*. Its pygidium and free cheeks remain unknown.

Jeffersonia Poulsen, 1927. – The type species was based on the pygidium of *J. exterminata* Poulsen from the Nunatumi Formation of northern Greenland. The name *Jeffersonia* had been used by E. O. Ulrich of the Smithsonian Institution, but only in manuscript. Ulrich had assigned several species from the Early Ordovician Jefferson City Formation of Missouri to his unpublished genus (hence the name: Ulrich's labels are still present in the Smithsonian Museum). These species were never published by Ulrich, but were eventually described from several localities, including Ulrich's, by Cullison (1944). Poulsen may have been using *Jeffersonia* in deference to Ulrich, but the outcome is that a very widespread clade from the later Ibexian is based on a pygidium alone. It is used again here, but with the caveat that new collections are made from the Nunatumi Formation may reveal the cephalic parts as incompatible with previous usage.

Licnocephala Ross, 1951. – The problems associated with this name have been discussed at length by Adrain & Westrop (2005). The type species *L. bicornuta* is known from silicified material from the western USA, but the material is small, and Ross only tentatively associated a pygidium. As employed herein, the characters of the pygidium are important in distinguishing the genus from *Chapmanopyge*. The concept of the genus is obviously vulnerable to

the proper description of the type species from new and better material.

Psephosthenaspis Whittington, 1953. – The type species was from an allochthonous boulder in Quebec, but description of three additional species from the base of the Whiterockian of Utah and Nevada by Fortey & Droser (1996) has identified the taxon as a good clade (also Adrain *et al.* 2012).

Rananasus Cullison, 1944. – From Missouri (and Oklahoma) was placed in the synonymy of *Bolbocephalus* by Adrain & Westrop (2005), but redescription of the type species, *R. conicus*, and of two other species, *R. brevicephalus* and *R. colossus* by Loch (2007), shows that the subdued, flat glabella is not convex (tr.) as it typically is in *Bolbocephalus* and suggests that together these three species may comprise a clade.

It is hardly possible to ignore the more problematic of these genera, because, as the cases of *Aponileus*, *Ceratopeltis*, *Gladiatoria* and *Psephosthenaspis* proved, they are capable of being clarified if new material becomes available. On the other hand, there are inevitable question marks over using a generic name based upon a fragmentary or poorly known type species. This problem recurs several times in the discussions below, and the general difficulties are outlined here so as to avoid too much repetition in the systematic section. Some of Poulsen's type species are from remote localities, but they are far easier to reach now than they were at the time the original types were collected. Several of the taxonomic problems might be resolved with the help of a focussed collecting trip. In general, we have employed existing generic names whenever possible, but the introduction of more genera in recent years means that it is difficult to continue to employ 'sensu lato' concepts in bathyurid taxonomy.

Morphology of Bathyuridae

A lot of new morphological information on bathyurids has been published in the last fifteen years, to say nothing of new (or clarified) genera. While J. M. Adrain and his colleagues are describing a wealth of silicified material of bathyurids comprised of dozens of species, with much new detail, from the Great Basin, western USA, it would be premature to produce a cladistic treatment of the whole family here. However, it may be useful to list some relatively neglected or misinterpreted characters that we consider will prove to be important in resolving clades in future work. These characters may prove to be more important in classification than the usual ones relating to glabellar structure, size of eyes and the like.

Genal structure. – The free cheeks of bathyurids are very diverse in morphology, and the homologies of structures developed there provide important characters. The lateral border furrow in most bathyurines is of usual trilobite type with the genal spine continuing it, but the round cross section may be modified to develop a blade-like profile. An inflated area may, however, develop close to the genal spine base, as in *Jeffersonia striagena* herein, or *Psephosthenaspis*. In bathyurellines, more complicated changes may occur. What we regard as the border furrow often curves across the cheek to approximately bisect it. The 'real' border then lies exterior to this furrow. This area may become variously inflated or prominent (e.g. *Punka flabelliformis*). But this furrow may also become effaced, in which case the genal field becomes uninterrupted and passes into an unusually broadly triangular genal spine, which frequently is orientated steeply in life position. This means that the broad genal area is a composite of true border plus genal field (see *Bathyurellus diclementsae* n. sp. and *Uromystrum affine* herein). To complicate the situation further, an entire border may be developed along the edge of the free cheek, as in *Uromystrum drepanon* n. sp. Only part of this is 'true' lateral border. More discussion of this feature is given in the descriptions of the species. Incorporating *Benthamaspis* into the family Bathyuridae allows for a contrary condition in which the entire adaxial part of the genal field is lost, and *all* the free cheek is effectively lateral border (more discussion, p. 60). The marginal rim that can develop on *Benthamaspis* is not homologous with the lateral border. Obviously, it is necessary to be particularly cautious in coding characters relating to genal borders, but they may well provide useful synapomorphies.

Cephalic doublure. – The width of the cephalic doublure relates to the border question just considered: it underlies 'true' border. However, in some taxa (*Uromystrum*, *Harlandaspis* gen. nov.), the doublure becomes tube like and recurved adaxially. This must have implications for the ventral structures: the condition may be associated with an elevated hypostome and illaenimorph anatomy. The inner margin of the doublure may show a vincular notch (*Harlandaspis*). The distribution of terrace ridges can also vary.

Hypostomes. – Relatively few hypostomes of bathyurids have been assigned (Whittington 1988), and attributions are always difficult to be certain about with fragmentary trilobites. Enough hypostomes are now known, however, to recognise that they are remarkably diverse within the family – more so, possibly, than in any other trilobite family, with the possible exception of Remopleurididae. Hypostomes of smaller Bathyurinae are relatively conservative proetoid in morphology: long (sag.) moderately convex middle body with posteriorly sited, straight middle furrows, surrounded by convex, but narrow borders. One like this is assigned to *Jeffersonia striagena* in this work. Larger-sized taxa show modifications from this pattern. Fortey (1979) assigned a triangular hypostome with a tumid middle body to *Petigurus*, and we have cautiously assigned a distinctive large, transverse hypostome with a subdued middle body to *Bolbocephalus* below (Fig. 6I–L). While these assignments are inevitably tentative, one of the advantages of having a specialised fauna is that there are very few candidates of the right size. Bathyurelline hypostomes include that here placed with *Uromystrum affine* and may show several peculiar features. Raised rims about the middle body are unusual, and the anterior wings are greatly extended dorsally (a modification probably connected with the specialisations of the cephalic doublure in this particular trilobite). A distinctive elongate hypostome is assigned with caution to *Harlandaspis* (Fig. 25AA). The hypostome assigned by Fortey (1979) to *Punka* is different again. If these assignments are confirmed, hypostome morphology should be useful in recognition of clades.

Pygidial borders. – The pygidial border in particular shows wide variation, especially in *Bathyurellus* and its relatives. The inner edge of the border may be sharply demarcated by a furrow, as in *Chapmanopyge*, or gradually sloping and indistinctly bounded. Pleural furrows may terminate, more or less abruptly, at the edge of the border. In some cases (see Fig. 27J), they deepen at this point. Or the furrows may continue on to the border with or without deflexion. One assumption might be that the well-furrowed border is the primitive condition. The basal bathyurid morphology shown by *Peltabellia* suggests the reverse: the border is initially without furrows (see Fig. 14H) and only secondarily acquires them. This is important in classification as it demonstrates that *Punka* species, for example, are united by this synapomorphy and may be part of a larger clade for which *Grinnellaspis* may eventually prove the senior name (Brett & Westrop

1996). *Uromystrum* has a typically concave border. There is also considerable variation in bathyurine pygidial margins, including border reduction and loss in *Petigurus*, and obliteration by effacement in some species of *Bolbocephalus* and *Rananasus*. However, some species of *Bolbocephalus* (e.g. *B. kindlei* Boyce, 1989; Fig. 8L) acquire defined interpleural furrows on the border region, as proved by the extent of the doublure beneath. Again, this is an advanced character.

Pygidial doublure. – There are interesting characters on the bathyurid pygidial doublure, not all of which have been considered in phylogeny. Although the inner doublure margin usually approximately parallels the pygidial margin, on *Bathyurellus* the inner margin has a v-shaped outline (Fig. 19T). The inner margin of the doublure usually turns dorsally, but in some species (e.g. *Peltabellia*, see Fortey & Peel 1990), there is a ventrally directed bar at the inner edge – an unusual feature in Trilobita. The deflexion dorsally of the doublure can be relatively gentle, or it can be very sharp (e.g. Fig. 27I), forming a wall at the posterior end of the body cavity. In some *Uromystrum* species, the doublure is so closely reflexed against the dorsal surface of the pygidium that they break out of the rock together (*U. affine* herein). In some cases, there is a match between cephalon and pygidium on the extent of the doublure and the definition of the their borders (*Licnocephala*, *Chapmanopyge*), but in other cases (*Uromystrum affine*), cephalic and pygidial borders are very different. All these features suggest adaptive significance attached to the doublure and the peripheral regions of the exoskeleton.

Systematic palaeontology

Remarks. – Except where stated, terminology follows the revised *Treatise on Invertebrate Paleontology* Part O (Whittington *et al.*, 1997). Within families, descriptions are arranged in such a way as to minimise repetition, so that for species that have been described previously, only a short, critical account is given. Holotypes are specified. All other specimens listed as 'Material' may be regarded as part of the type series. Specimens were prepared manually using a vibro tool and needles, but a few could not be completely exposed due to silicification or fragility. Open nomenclature is employed for many imperfectly known species.

Repository. – Type and figured specimens are curated in the Geological Museum (now incorporat-

ing the former Palaeontologisk Museum), hereafter abbreviated PMO) under the Naturhistorisk Museum, University of Olso or the Sedgwick Museum, Cambridge (CUSM), with a small duplicate collection in the Natural History Museum, London. Comparative material is illustrated from the collections of the US National Museum, Washington D.C. (USNM), and the Peabody Museum, Yale University (YPM).

Order Aulacopleurida Adrain, 2011
Family Hystricuridae Hupé, 1953

Discussion. – Although they have been regarded as a sub-family of the Solenopleuridae, the hystricurids underwent an evolutionary radiation in the palaeotropical regions of the Early Ordovician, and it is likely that they comprise a true clade. The classification of 'generalised' ptychoparioids is still uncertain, and for this reason, the Hystricuridae is treated as a separate family herein, as did Adrain *et al.* (2003).

Genus *Svalbardicurus* n. gen.

Type species. – *Svalbardicurus delicatus* n. sp.

Etymology. – From Svalbard + the ending of the related *Hystricurus*.

Diagnosis. – Dorsally smooth hystricurids with short pre-glabellar field; palpebral lobes weakly curved and relatively far forwards; sub-triangular pygidium with wide, blunt-tipped axis extending far posteriorly having three axial rings and large terminal piece, posterior border steep, wall like; pygidial interpleural furrows absent.

Discussion. – Adrain *et al.* (2003) have brought new criteria to bear on hystricurid classification resulting in the creation of a number of new genera, which are finely divided one from another. The species described below might formerly have been included in *Paraplethopeltis* Bridge & Cloud, 1947, along with a number of other dorsally smooth hystricurids. Hintze (1953) included a number of species (with question) from the Ibexian of Utah in this genus, and Boyce (1989) followed suit when he assigned *Bathyurus seelyi* Whitfield, 1889 to *Paraplethopeltis*, identifying new specimens from western Newfoundland with the types from New York State. The latter species is clearly closely related to the new one described below. The type species of *Paraplethopeltis*, *P. obesa*

Bridge & Cloud, 1947, and two other species attributed by them to the same genus, differ from both *B. seelyi* and the Spitsbergen species in lacking a cranidial border and strongly defined occipital ring and in having smaller palpebral lobes; the pygidium is of the generalised kind more typical of early and primitive hystricurids and of some plethopeltids, having pleural and interpleural furrows and a narrow border. In these respects, *Paraplethopeltis* is clearly different from the species from New York State, Newfoundland and Spitsbergen. The lack of surface ornament is the main feature shared with *Paraplethopeltis*, a feature of relatively little importance; on the other hand, the wide pygidium with its steep posterior margin is a good apomorphic feature of the species from Spitsbergen and eastern USA and Canada, for which we propose the new generic name *Svalbardicurus*. It seems likely that *Hystricurus politus* Poulsen, 1937, from East Greenland, and '*Hystricurus*' sp. of Landing *et al.* (2012) and *Bathyurus seelyi* Whitfield, 1889 from northeastern USA are congeneric.

The closest related genus to *Svalbardicurus* is probably *Hystricurus sensu stricto*, as revised by Adrain *et al.* (2003), which also has large eyes, long (tr.) post-ocular cheeks, a border about as wide as the pre-glabellar field and an 'advanced' hystricurid pygidium with (up to) four ribs that have a tendency to be prolonged into short spines and a steep border. *Hystricurus* species are usually coarsely tuberculate, have strongly curved palpebral lobes and narrow (exag.) post-ocular cheeks; their pygidia retain interpleural furrows, and the terminal piece on the axis is often bilobate. *Politicurus* Adrain *et al.* 2003 also has subdued sculpture, but differs from *Svalbardicurus* in its very posterior eyes, occipital spine and primitive – style pygidium.

Svalbardicurus delicatus n. sp.

Figure 4

Holotype. – Pygidium, PMO 208.183 (Fig. 4L, M); Spora Member.

Material. – Cranidia, PMO 208.065-6, 208.180, 208.182, CAMSM X 50188.066, CAMSM X 50188.069; free cheek, CAMSM X 50188.65; pygidia, PMO 208.061, CAMSM X 50188.64, X 50188.70.

Stratigraphical range. – Spora Member of the Kirtonryggen Formation (undivided), early Ibexian (Stairsian), *Svalbardicurus delicatus* fauna.

Diagnosis. – *Svalbardicurus* with relatively long glabella and gently convex border; pygidium with median axial tubercles and weakly spinose ends to pleural ribs.

Etymology. – Latin 'delicate' when compared with coarsely tuberculate species.

Description. – Most complete cranidium is about three-quarters as long as wide (across posterior limbs), while the width at the palpebral lobes is less than the length. Glabella gently convex sided and gently forwardly tapering to somewhat truncate-rounded front. Glabella evenly convex (tr.) and occupies about 85% cranidial length and completely lacks lateral glabellar furrows, but the occipital ring is well defined and one-sixth glabellar length. Axial furrows narrow and deep; pre-glabellar furrow hardly shallower. Occipital ring of even width along its length, without prominent median tubercle. Midpoint of palpebral lobes close to mid-length of cranidium; palpebral furrow nearly parallel to axial furrow, such that the narrow (tr.) lobes are hardly curved. Lobes are one-third glabellar length (exsag.). Very faint eye ridges. Posterior section of facial suture diverges outwards at about 60° to sag. line to cut posterior border at an acute angle, thereby defining a rather long (exsag.) post-ocular cheek. Anterior section very slightly divergent in front of eyes before curving sharply adaxially anteriorly. Posterior border furrow deep, arched slightly forwards, marking off narrow but convex posterior border. Steeply downsloping pre-glabellar field slightly longer (sag.) than mid-part of anterior border in dorsal view. Deep border furrow curved slightly backwards medially, where the gently convex anterior border is widest. Free cheek known from incomplete example, showing convex lateral border extending into a short, triangular genal spine.

Pygidium sub-triangular, twice as wide as long, axis wider (tr.) than pleural fields at their widest. Pleural fields are nearly horizontal, while the posterior margin is sub-vertical. Axis 80% as wide as long, convex and tapering gently to the posterior pygidial margin; three axial rings of decreasing width and length, defined by ring furrows that shallow laterally, most noticeably on the third. Terminal piece three times wider than long. Median tubercles present towards anterior edge of all three rings and the terminal piece – this probably indicates that there is a fourth effaced ring incorporated into the latter. Four pleural ribs increasingly backwardly curved posteriorly, of which the fourth is a drop-like lobe. Interpleural furrows absent. Tips of ribs extended into blunt spines that appear to be variable – the anterior

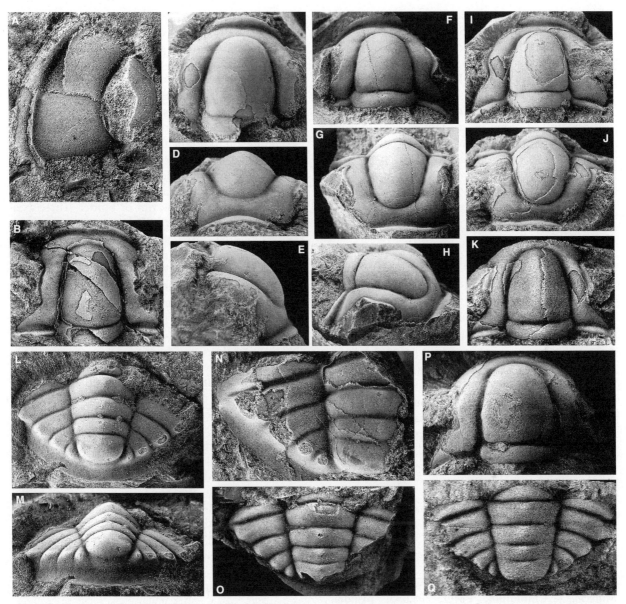

Fig. 4. *Svalbardicurus delicatus* n. gen., n. sp. All Spora Member (1–20 m, undivided horizons). **A**, Free cheek CAMSM X 50188.65 in lateral view (x6). **B**, Cranidium with relatively long glabella CAMSM X 50188.66 in dorsal view (x3). **C–E**, Cranidium PMO 208.066 in dorsal, anterior and lateral views (x5). **F–H**, Cranidium PMO 208.182 in dorsal, anterior and lateral views (x3). **I–J**, Cranidium PMO 208.065 in dorsal and anterior views (x3). **K**, Cranidium CAMSM X 50188.69 in dorsal view (x5). **L–M**, Holotype, pygidium PMO 208.183 in dorsal and posterior views (x5). **N**, Incomplete pygidium with subdued axial tubercles PMO 208.061 in dorsal view (x6). **O**, Pygidium CAMSM X 50188.64 in dorsal view (x5). **P**, Cranidium PMO 208.180 in dorsal view (x5). **Q**, Pygidium CAMSM X 50188.70 in dorsal view (x5).

rib carries quite a prominent spine on one specimen. Facet extends across more than half pleural width. Posterior border forms a near-vertical 'wall' around margin.

Surface of exoskeleton lacks any surface sculpture.

Discussion. – Boyce (1989) reillustrated the New York State types of *Bathyurus seelyi* Whitfield, 1889, together with other specimens from western Newfoundland he regarded as conspecific. They are closely similar to *Svalbardicurus delicatus* and must be assigned to the same genus. Cranidia are apparently identical; however, pygidia of *S. seelyi* are less transverse and lack median tubercles and the spinose tips of the pleural ribs. Boyce also illustrated the cranidium of *Paraplethopeltis cordai* (Billings, 1859) from the Levis conglomerate, Quebec, which again is similar to that of *S. delicatus*, apart from a more tapering glabella and convex anterior border. Cranidia of '*Hystricurus*' sp. from the Rochdale

Formation illustrated in Landing *et al.* (2012) are also similar to those of the Spitsbergen species, but a fine surface sculpture of ridges distinguishes them. *Hystricurus nudus* Poulsen, 1937, from northeast Greenland, known from an incomplete cranidium, is also smooth, but has been shown by McCobb *et al.* (in press) to belong to *Millardicurus*. *Svalbardicurus* was widespread along the eastern margin of Early Ordovician Laurentia, although rarely collected. In the concept of Adrain *et al.* (2003), *Svalbardicurus* would belong to the sub-family Hystricurinae and is not related to another genus with reduced surface sculpture, *Politicurus* Adrain *et al.* (2003) in the sub-family Hintzecurinae.

Genus *Hystricurus* Raymond, 1913

Type species. – *Bathyurus conicus* Billings, 1859, Beauharnois Formation, Montreal, Quebec.

Remarks. – The type specimen of the type species was illustrated by Adrain *et al.* (2003); additional material was attributed to it by Desbiens *et al.* (1996). We accept the redefinition of the genus given by Adrain *et al.* (2003), to accommodate hystricurids with long palpebral lobes extending far forwards and comparatively large, well-segmented pygidia.

Hystricurus cf. *Hystricurus* sp. nov. B Adrain *et al.*, 2003

Figure 5A–D, H–M

Material. – Cranidium, PMO 208.055; pygidia, PMO 208.056-7.

Stratigraphic range. – Spora Member (undivided), Kirtonryggen Formation, early Ibexian (Stairsian), *Svalbardicurus delicatus* Fauna.

Description. – Cranidium nearly twice as wide at the posterior margin as long (sag.) in dorsal view, just less than a centimetre in length. Parabolic glabella widest at occipital ring, in front of which axial furrows are bowed inwards. Oval smooth area adjacent to axial furrow opposite posterior part of palpebral lobe is only indication of glabellar segmentation. Very long palpebral lobes are 60% or more glabellar length and extend not quite to front of glabella, but gently curved, such that the interocular cheek is about twice as wide medially as at its rear

end. Steep downward slope of pre-glabellar field and adjacent pre-ocular cheek stops at deep border furrow. Anterior border gently bowed forwards and smooth, weakly convex. Eye extends so far back that the posterior branch of the facial suture curves very slightly forwards behind the eye, before curving distally backwards. Posterior border furrow deep. Surface sculpture of rather fine, scattered tubercles, extending on to pre-ocular cheeks.

Pygidium has similar tubercles on axis and front part only of anterior pleurae. It is convex, with lateral parts of pleural fields nearly vertical until reaching a prominent narrow rim or border. Three axial rings are obvious; a fourth if present is weak at most. Corresponding three pleural segments with deep pleural furrows, and only the anterior interpleural furrow, and that weak and adaxially developed. Fourth pleural furrow short, running exsagitally and only visible in posterior view. The posterolateral tip of the first pleural lobe is inflated. There is variation in the width of the posterior border, but it is probably intraspecific.

Discussion. – Although well preserved, the material of this species is not sufficient to name it formally. It conforms to the definition of *Hystricurus s. s.* in Adrain *et al.*, 2003. The palpebral lobes are longer than in any other species. The type species has shorter palpebral lobes and coarser tuberculation, as do the species described in Boyce (1989). Adrain *et al.* (2003) figured four Stairsian species from the Ibexian of the Great Basin, and of these, the species from the Spora Member is strikingly like their *Hystricurus* sp. nov. B., with palpebral lobes long enough to proximally constrict the posterior limb of the fixed cheek. Details of ornament agree well. The formal taxonomy of the species from the western USA is pending, and beautifully preserved, silicified material obviously provides the best basis for a species name, hence our open nomenclature. The only significant difference is the wider pygidium of the USA material, with suggestions of median axial tubercles, and the single pair of anterolateral pleural knobs are more pronounced. It is likely that they are similar in age.

We have another larger pygidium (PMO 221.247) from the Spora Member (Fig. 5E–G) with prominent median axial tubercles. It lacks surface sculpture other than fine granulation and has a sloping border. It is probably the pygidium of another hystricurid and the differences it displays from the species under discussion are very likely too numerous to be accountable by ontogeny. It is referred to as 'hystricurid gen. and sp. indet.' on the figure descriptions.

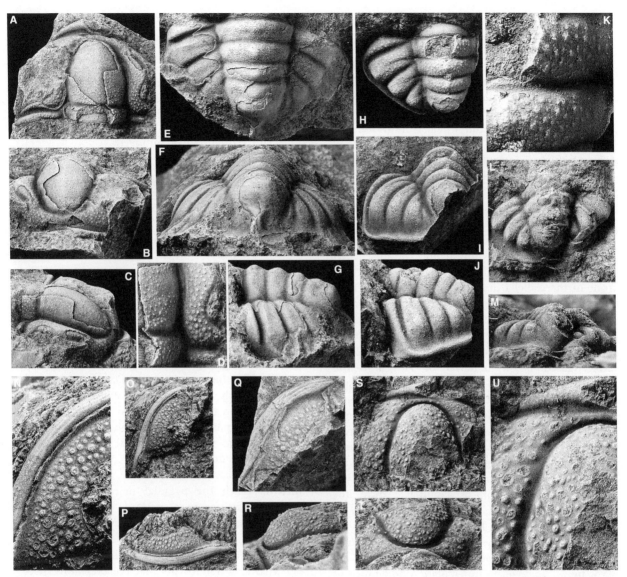

Fig. 5. **A–D, H–M,** *Hystricurus* cf. *Hystricurus* sp. nov. B Adrain *et al.*, 2003. Spora Member (1–20 m, undivided horizons). **A–C,** Cranidium PMO 208.055 in dorsal, anterior and lateral views (x3). **D,** Enlargement of A showing the sculpture on glabella and fixed cheek (x6). **H–J,** Pygidium PMO 208.057 in dorsal, posterior and lateral views (x5). **K, L,** Pygidium PMO 208.056 in dorsal and posterior views (x5). **M,** Enlargement of H showing the sculpture on pygidial axial rings. **E–G,** Hystricurid gen. et sp. indet. Spora Member. Pygidium PMO 221.247 in dorsal, posterior and lateral views (x3). **N–U,** *Hystricurus* sp. 1. Spora Member. **O–P,** Free cheek PMO 221.249 in dorsal and lateral views (x2). **N,** Enlargement of O showing the sculpture (x4). **Q,** Free cheek PMO 221.248 in dorsal view (x2). **R–T,** Incomplete cranidium PMO 221.245 in lateral, dorsal and anterior views (x2). **U,** Enlargement of S showing the coarse sculpture.

Hystricurus sp. 1

Figure 5N–U

?1989 *Hystricurus oculilunatus* Ross; Boyce (*pars*), pl. 8, figs 5–8, pls 9–11, *non* pl. 8, figs 1–4.

Material. – Cranidium, PMO 221.245; free cheeks, PMO 221.248-9.

Stratigraphic range. – Spora Member (undivided) of the Kirtonryggen Formation.

Discussion. – Boyce (1989) figured and described two very coarsely tuberculate species from western Newfoundland that can be compared with a second *Hystricurus* species from the Spora Member. One he named as *H. deflectus* Heller, 1954, but without revising the original type material. This hystricurid species has tuberculation even coarser than that of the type of *Hystricurus conicus* (Adrain *et al.* 2003, fig. 3), and the palpebral lobes are also much shorter than the species described above. A second, coarsely tuberculate species was identified by Boyce with *H. oculilunatus* Ross, 1951, a species originally described from the western USA. Boyce also figured material similar to that from Newfoundland from the Fort Ann Formation, Plattsburgh, New York. The latter is now known as the Rochdale Formation, from which Adrain and Westrop (*in* Landing *et al.* 2012) redescribed *Hystricurus crotalifrons* (Dwight, 1884), a species with remarkably coarse tuberculation. We have an incomplete, very large cranidium and several free cheeks from the Spora Member that are generally similar to the material described by Boyce (1989) from the Lower Member of the Boat Harbour Formation in western Newfoundland. The absence of a median furrow bisecting the pre-glabellar field in front of the glabella distinguishes this species from *H. deflectus*. One of our free cheeks retains enough cuticle to see very striking parallel raised ridges on the border, which are not reflected on internal moulds. However, now that Adrain *et al.* (2003) have figured larger material of *H. oculilunatus* from the type area, the identification with the species from Newfoundland and New York seems more questionable, since the latter has a differently shaped glabella and altogether coarser and different ornament and pygidial details. The poor material available from the Spora Member cannot provide the basis for a new name. The revision of the type species, *Hystricurus conicus* (Billings), from the Beauharnois Formation, Quebec, by Desbiens *et al.* (1996) is also relevant, because this species is very much like Boyce's (1989) Newfoundland 'oculilunatus' and includes some morphs (Desbiens *et al.*, pl. 3, fig. 8) as coarsely tuberculate – and more so than Billings' type specimen. However, the stratigraphic occurrence of *H. conicus* is Blackhillsian (= Cassinian), which is surprisingly younger than any hystricurids in Newfoundland and Spitsbergen. There is a case for revising the whole group of *Hystricurus s.s.* that resemble the type species, and for the present, we name our material under open nomenclature.

Order Proetida Fortey & Owens, 1975
Family Bathyuridae Walcott, 1886
Sub-family Bathyurinae Walcott, 1886

Remarks. – This sub-family is used here while recognising that it is a paraphyletic assemblage of genera, comprising all remaining bathyurids with the clade Bathyurellinae removed.

Genus *Bolbocephalus* Whitfield, 1890

Type species. – *Bathyurus seelyi* Whitfield, 1886, Fort Cassin Formation, New York State. See Whittington (1953).

Discussion. – Stratigraphically early species of *Bolbocephalus* have glabellas that expand forward in a regular way, while some later species develop a more conspicuous posterior constriction in front of the occipital ring, and a more inflated frontal glabellar lobe, giving the axial furrows a sigmoidal profile. The former resemble *Petigurus* more closely, to which *Bolbocephalus* is allied. Variation in pygidial morphology relates particularly to the development of the interpleural furrows and the border, as it does in several other bathyurids. The extreme development is seen in *Bolbocephalus kindlei* Boyce, 1989, on which deep interpleural furrows reach the pygidial margin; however, no cephalic parts were associated with this species. Some species have no indication of furrows on the border, while others, like the type species, have interpleural furrows that may extend on to the border. It is also noteworthy that *Bolbocephalus* species probably attain the largest size of any bathyurid. Pygidia may approach 7 cm in width; it is plausible that whole trilobites were at least twice that length. The assumption that bathyurids were always small- or medium-sized trilobites is incorrect. Both *Petigurus* and *Bolbocephalus* are more phacomorph than other members of the family, with their expanded anterior glabellar lobes, although coarse

Fig. 6. Bolbocephalus gunnari n. sp. Lower part of Nordporten Member, *P. groenlandicus* fauna. **A–C,** Cranidium PMO 208.114 in dorsal, lateral and anterior views (x5), 105 m from base of member. **D,** Enlargement of N showing structure. **E–F,** Free cheek PMO 208.068 in dorsal (x3.5) and lateral (x5) views, the latter showing sculpture, 105 m. **G,** Small cranidium PMO 223.162 in dorsal view (x8), 105 m. **H,** Pygidium PMO 208.241, basal bed, in dorsal view (x6). **I–J,** Hypostome, incomplete and cautiously associated PMO 223.251 in dorsal and posterior views (x4), 105 m. **K–L,** Hypostome cautiously associated PMO 223.181 in dorsal and lateral views (x3), 105 m. **M,** Internal mould of free cheek PMO 208.112 in lateral view (x2), 105 m. **N–P,** Holotype, pygidium PMO 223.187 in dorsal, posterior and lateral views (x6), 105 m. **Q–S,** Small pygidium PMO 223.161 in dorsal, posterior and lateral views (x7), 105 m. **T,** Pygidium PMO 208.071 in dorsal view (x3), 105 m. **U,** Pygidium with relatively effaced axis PMO 223.232 in dorsal view (x4), 105 m. **V–W,** Large pygidium PMO 208.260 in dorsal and posterior views (x2), H collection, low in *P. groenlandicus* fauna, exact horizon unknown.

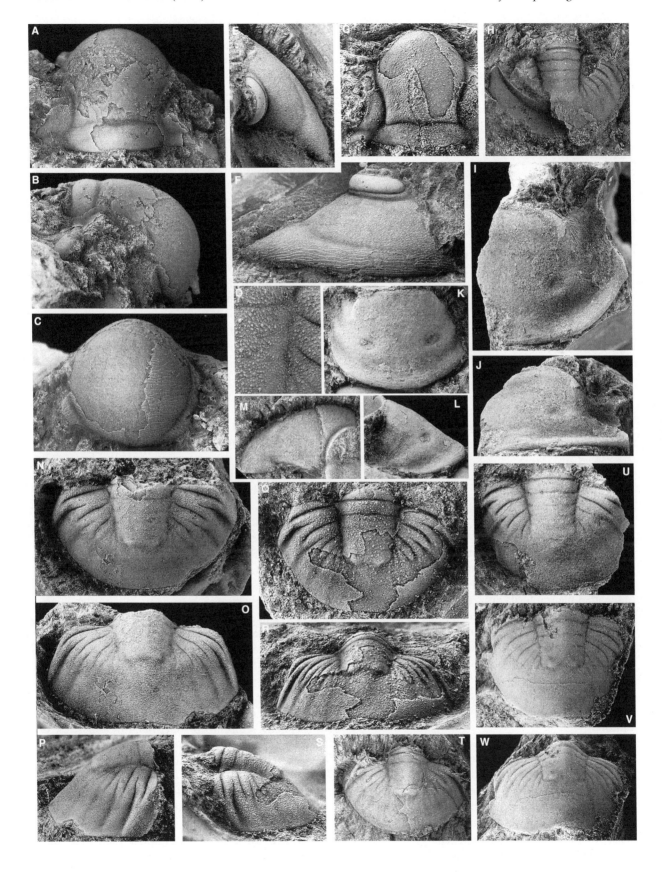

sculpture of the appropriate type is present only on *Petigurus*, but not on any described *Bolbocephalus* species. As noted below (p. 54), the type species of *Gelasinocephalus*. *Gelasinocephalus whittingtoni* Loch, 2007, is very much like *Bolbocephalus*, but does have a relatively long (sag.) pre-glabellar field. This may be a plesiomorphic character in Bathyuridae and therefore not of significance in classification above the species level.

Bolbocephalus gunnari n. sp.

Figure 6

2007 *Bolbocephalus* sp. 2. Loch, p. 28, plate 7, figs 3, 4.

?2007 *Bolbocephalus* sp. Loch, p. 29, plate 7, figs 16, 17.

Holotype. – Pygidium, PMO 223.187 (Fig. 6N–P); 105 m from base of Nordporten Member.

Material. – Cranidia, PMO 208.114, 223.162, 208.139; free cheeks, PMO 208.068, 208.112; pygidia, PMO 208.071, 208.104, 208.241, 208.260, 223.161, 223.232, 223.234, CAMSM X 50188.59; hypostome, PMO 223.181, 223.251.

Stratigraphic range. – Nordporten Member, 0–105 m (?130 m), *Petigurus groenlandicus* fauna.

Diagnosis. – *Bolbocephalus* species with pygidial interpleural furrows distinct but weak on exterior part of pygidial pleural fields only, and extending to margin; pygyidial axial rings comparatively poorly defined. Blunt genal spine. Surface sculpture characteristic, over much of dorsal surface: very fine tubercles or granules that tend to coalesce into low ridges.

Etymology. – For Gunnar Henningsmoen, a good friend and scientist who shared fieldwork with the authors.

Description. – Cranidium very convex (sag.) with inflated and expanded frontal glabellar lobe that overhangs border, in profile making a semicircle. In dorsal view, width of the glabella (tr.) at its widest anteriorly is similar to the length along the sagittal line and about the same as the transverse width of the occipital ring. Narrow axial furrows bowed broadly inwards in front of occipital furrow, then curving outwards to outline the inflated frontal gla-

bellar lobe. No lateral glabellar furrows defined. In anterior view, the front of the glabella makes an almost perfect circle. Occipital furrow moderately deep and transverse, defining occipital ring about 25% glabellar length in dorsal view, widest medially and tapering anteriorly towards glabella constriction, Pre-ocular fixed cheeks narrow (tr.) and downsloping, while post-ocular cheeks flare outwards in typical bathyurid fashion. Palpebral lobes medially positioned, well defined, gently arcuate in outline.

Free cheek relatively long and narrowly triangular, with convex pleural field, and border continuing its downward slope. Border furrow shallow, especially anteriorly, and curving adaxially at genal spine. Border widens towards stout, bluntly triangular genal spine, which seems to be proportionately larger on big specimens. Eye socle prominent, convex, described by distinct furrow above and below, and slightly wider anteriorly. Curved eye is three times longer than high, with minute holochroal lenses.

We describe an internal mould of a hypostome here (Fig. 6K), although there is no final way of proving whether it belongs to *B. gunnari* or *B. convexus*. It is large enough to be that of *Bolbocephalus*; a posteromedian point is lacking, and it is generally transverse and flat compared with the hypostome assigned to *Petigurus* by Fortey (1979) and hence unlikely to be the hypostome of appropriately sized *P. convexus*. Middle body is weakly defined and hardly inflated and merges somewhat with lateral border. Anterior margin is almost transverse, slightly convex. Anterior wings are hardly developed (contrast *Uromystrum affine*). Middle furrows are two oval depressions, slightly inflated margins behind them. Slightly convex, weakly arcuate border carries terrace ridges running parallel to margin. In that these are reflected on the internal mould, which could be taken as evidence that this hypostome belongs rather to *B. convexus* with its stronger ridges, but since this is the first attribution of a hypostome to *Bolbocephalus*, we are uncertain of the significance of this feature.

Cephalic sculpture of fine, but low tubercles, or flattened granules, which amalgamate anteriorly into more or less transverse ridge-like structures. Wavy fine ridges are carried on the dorsal surface of the border, running more or less parallel with the edge of the cheek. This sculpture is not reflected on the internal mould.

Pygidium 1.45–1.65 times wider than long, axis extending to 0.6 pygidial length (dorsal view), with maximum transverse width at back end of facets. Maximum width of pleural field is about the same (tr.) as that of the axis at first ring. Axis hardly longer than the width of anterior axial ring, very gently tapering to more or less squared-off terminal piece. First ring is always comparatively well defined, but posterior two or three rings weakly so, or not at all. Articulating half-ring is almost as wide (sag.) as anterior ring. Where dorsal surface is preserved (Fig. 6D, N), paired tubercles at mid-ring level to either side of the sag. line reveal segmentation: there are five pairs of such tubercles, presumably corresponding to the usual four bathyurid segments plus a fifth on the terminal piece, which can otherwise not be recognised other than as a swollen area, usually subtly bilobate. A second series of tubercles exterior to the first may be present. Much of lateral and posterior parts of pleural field slope down straight to the margin, with sub-horizontal adaxial parts tapering posteriorly to end of axis. Pleural furrows extend only to top part of slope; three pairs clear and fourth pair obscure, more steeply backward sloping. Interpleural furrows begin towards ends of pleural furrows so that they overlap only along the steep edge of the pleural fields. However, the first two pairs of interpleural furrows, and variably the third, continue downwards towards the pygidial margin which they very nearly reach, the furrows becoming wider and more diffuse in the process. Facet steep, triangular, slightly dished. Posterior margin of pygidium shows gentle posteromedian arch. Doublure sub-horizontal and then steeply reflexed with a groove (ridge ventrally) at the change in slope, extending up to the axial tip, where the vertical 'wall' is highest. Sculpture also of flattened tiny tubercles, with a tendency to line up as ridges. The smaller pygidium (Fig. 6Q) shows paired larger tubercles also on the proximal part of the pleural fields.

Discussion. – Among *Bolbocephalus* species, the structure of the pygidium of *B. gunnari* is distinctive, with the interpleural furrows developed on the flanks only, a feature which readily distinguishes the species from others in the Kirtonryggen Formation. The type species (Whittington 1953) has longer furrows extending adaxially. *Bolbocephalus angustisculatus* Poulsen, 1937, is known from a fragment of a pygidium, but it does show interpleural furrows extending on to the proximal part of the pleural field, but not on the border – the opposite of *B. gunnari*. *B. kindlei* Boyce, 1989, has altogether much deeper interpleural furrows, and the posterior pair extends clearly and deeply to the pygidial margin. The surface sculpture of *B. gunnari* is definitive, although material of some previously described species is insufficiently well preserved to see sculpture.

A pygidium very similar to that of *B. gunnari*, showing the paired tubercles on the axis as well as the characteristic pleural furrows, was described from the Kindblade Formation of Oklahoma by Loch (2007) as *Bolbocephalus* sp. 2. These distinctive shared characters are considered taxonomically more important than the slightly narrower axis of the Oklahoma specimen. The specimen illustrated was derived from Loch's *Benthamaspis rhochmotis* Zone, but Loch (2007 p. 29) states that the species extends into the succeeding zone. We believe it probable that the same species is present in Oklahoma as in Spitsbergen.

The species is found through a considerable thickness of strata in Spitsbergen. A large pygidium from the locality H (isolated limestone we consider to be about 100 m from base of Nordporten Member) has weaker interpleural furrows distally, and the paired axial tubercles are not present at this size, at least on the internal mould. Such differences as there are from the holotype are attributed to ontogenetic change. It is possible that a large pygidium figured (under open nomenclature) by Loch (2007, pl. 7, figs 16, 17) represents the same stage of growth, especially as it has an arch on its posterior margin. Pygidia from the base of the range of *B. gunnari* (Fig. 8V) are apparently identical to those from the top of its range.

Bolbocephalus convexus (Billings, 1865)

Figure 7

1865 *Dolichometopus?* *convexus* n. sp. Billings, p. 269, fig. 253.

?1937 *Bolbocephalus groenlandicus* sp. nov. Poulsen, pp. 48–49, pl. 5, figs 9–13.

?1979 *Bolbocephalus convexus* (Billings); Ludvigsen, p. 861, pl. 1, figs 5, 6.

1979 *Bolbocephalus convexus* (Billings, 1865); Fortey, pp. 78–80, pl. 26, pl. 28, figs 9, 12.

1989 *Bolbocephalus convexus* (Billings, 1865); Boyce, pp. 46–47, pl. 22, figs 1–8.

Material. – Cranidia, PMO 223.163, 223.189, 208.283; free cheek, PMO 208.259, 223.182; pygidia, PMO 223.209, 223.170, PMO NF 1907, CAMSM X 50188.40.

Stratigraphic range. – Nordporten Member 105–200 m from base; associated with upper part of the *Petigurus groenlandicus* fauna and *P. nero* fauna.

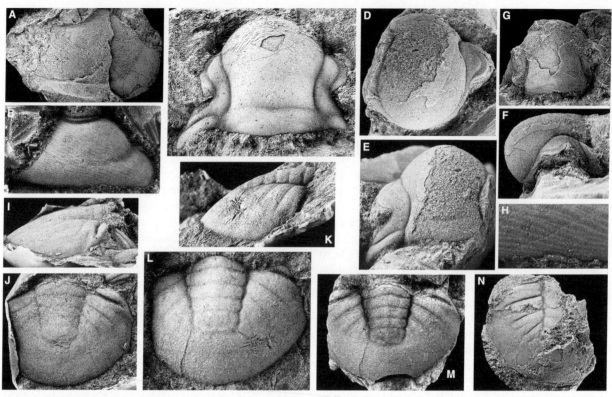

Fig. 7. Bolbocephalus convexus (Billings, 1865). Nordporten Member, *P. groenlandicus – P. nero* faunas. **A**, Large free cheek PMO 208.259 in lateral view (x0.5), **H** collection, low in *P. groenlandicus* fauna, exact horizon unknown. **B**, Free cheek showing sculpture PMO 223.182 in lateral view, (x2.5), 105 m. **C**, Cranidium PMO 223.163 in dorsal view (x6), 105 m. **D–F**, Cranidium PMO 223.189 in dorsal, anterior and lateral views (x4), 105 m. **G**, Cranidium PMO 208.283 in dorsal view (x3), **H**, collection. Enlargement of border of specimen in **B** showing raised ridges. **I–J**, Latex cast of pygidium PMO 223.209 in lateral and dorsal views (x2), 105 m. **K–L**, Pygidium PMO 223.170 in lateral and dorsal views (x4), 105 m. **M**, Pygidium PMO NF 1907 in dorsal view (x3), precise horizon unknown. **N**, Very large fragmentary pygidium CAMSM X 50188.40 in dorsal view (x0.8), ca. 130 m, with faint interpleural furrows.

Diagnosis. – *Bolbocephalus* with pygidium lacking interpleural furrows, and pleural furrows weak and straight and not passing on to border region. Short (sag.) glabella with broadly rounded frontal lobe. Cephalic surface sculpture of strong terrace lines on glabella and free cheeks, but apparently not on pygidial dorsal surface.

Discussion. – The type specimen from western Newfoundland was illustrated by Fortey (1979, pl. 26, fig. 8) with other material assigned, and the general description given in that work does not need repetition. It was found with the *Petigurus nero* fauna in the Catoche Formation, where other species of Billings (1865) occur. Boyce (1989, fig. 4) further extended the range of *Bolbocephalus convexus* downwards into the underlying fauna of the Barbace Cove Member at the top of the Boat Harbour Formation. This is probably into strata equivalent to that yielding the *Petigurus groenlandicus* fauna in Spitsbergen. Since none of the other species described in this work are the same between the two *Petigurus* faunas, it was a matter of interest whether the early and late

representatives of *Bolbocephalus convexus* were or were not similar. Boyce (1989) also placed *B. groenlandicus* Poulsen, 1937, into synonymy with *B. convexus*, the former being a species recorded alongside some of the other characteristic trilobites of the *Petigurus groenlandicus* fauna. The new material herein generally confirms Boyce's interpretation. Boyce did not illustrate a free cheek for his stratigraphically early material, but one from the Nordporten Member shows the strong terrace ridges that, unusually, pass obliquely across the pleural field and straight on to the rounded border (Fig. 7B, H). The same feature is seen on the free cheek of the younger specimen from the Catoche Formation figured by Fortey (1979, pl. 26, fig. 9). Terrace ridges are somewhat finer on the latter than the former. Similar terrace ridges are seen on the (incomplete) glabellas of *Bolbocephalus groenlandicus* Poulsen; they are strong enough to be reflected on the internal moulds in front of the occipital ring. However, we have an incomplete cranidium from 105 m above the base of the Nordporten Member (Fig. 7C) with well-preserved cuticle that shows the terrace ridges only on the anterior part of the glabella, where on other material they are apparently present further back. At present, we assume that this is a variant within the species. As far as pygidia are concerned, there is variation in overall length/width ratios (dorsal view) and in expression of the two or three pairs of short, adaxial pleural furrows. The pygidium attributed to *B. groenlandicus* is 75% as long as wide; the most transverse specimen is possibly that of Fortey (1979, pl. 26, fig. 2), which is about 57%; the same ratio on the type is 60%. Other stratigraphically early specimens yield values about 65%, so there is not enough systematic variation to suggest a specific difference. There also seems to be some variation in the clarity of both pygidial axial rings and pleural furrows, but various degrees of effacement are shown which suggest variations in preservation superimposed on minor original differences. Hence, *B. convexus* is regarded at the moment as the most long ranging of the species in the Kirtonryggen Formation. *Bolbocephalus stitti* Loch, 2007, from the Kindblade Formation of Oklahoma, has a generally similar pygidium to that of *B. convexus*. Its cranidium is less convex (sag.), and the free cheek has a deeply defined lateral border, to which the terrace ridge sculpture appears to be confined.

Bolbocephalus stclairi Cullison, 1944

Figure 8A–I

1944 *Bolbocephalus st. clairi* n. sp. Cullison, pp. 78–79, pl. 35, figs 27–31.

2007 *Bolbocephalus sancticlairi* nom. nov. Loch, p. 23.

2007 *Bolbocephalus* sp. Loch, pl. 7, fig 8.

Material. – Cranidia, PMO 223.174, CAMSM X 50188.42; pygidia, PMO 208.201, 208.090.

Stratigraphic range. – 105–190 m from base of Nordporten Member, upper *Petigurus groenlandicus* fauna to *P. nero* fauna.

Diagnosis. – *Bolbocephalus* with elongate glabella, in dorsal view glabellar length 1.3–1.4 times maximum width across frontal lobe, very slightly waisted in front of occipital ring, and with low convexity (sag.) except at frontal glabellar lobe. Short pre-glabellar field; adaxial part of posterior fixed cheek inflated. Pygidium with three or four

pairs of narrow pleural furrows and interpleural furrows hardly defined.

Discussion. – This species was named by Cullison, 1944, as *Bolbocephalus st. clairi*, a name that unfortunately already existed as a *nomen nudum* coined by E. O. Ulrich, in Weller & St Clair (1928). Judging that name not available for use by Cullison (1944), Loch (2007) proposed *B. sancticlairi* as a replacement name. This was not necessary as Cullison (1944) subsequently named and validated the species name *st clairi*, based upon the same material as the original *nom. nud.* The low curvature of the glabellar profile, mostly confined to the anterior part in lateral view, coupled with the relatively slight constriction in transverse width of the glabella at a level behind the palpebral lobes serve to distinguish it from most other species of *Bolbocephalus*. We can find no differences between Cullison's holotype cranidium and those from Spitsbergen other than a slightly more rounded anterolateral profile of the glabella in the latter in dorsal view. The stratigraphically early species *B. jeffersonensis* Cullison, 1944, is the only species with virtually no constriction of the glabella, which may be taken as the primitive condition. A cranidium figured under open nomenclature by Loch (2007, pl. 7, fig. 8) is apparently identical to our material of *B. stclairi*. The most similar other named species is probably *B. myktos* Loch, 2007, from the earlier part of the Kindblade Formation, but the one cranidium illustrated of this species shows a more forwardly protruding glabella and an acuminate, as opposed to rounded glabellar front. It is also small compared with the Spitsbergen material. The pygidium attributed to *B. myktos* is unlike that of other *Bolbocephalus* spp. in having a transverse posterior margin, and no pygidium like it has been found in Spitsbergen (it resembles that of *Gelasinocephalus whittingtoni* Loch, 2007). However, a pygidium of *Bolbocephalus* from our collections does seem a good match for that of *B. stclairi* (Fig. 8G), having a compatible convexity, relatively low axis and weakly defined, but long pleural furrows that continue close to the pygidial margin. No evidence of surface sculpture is available on any of our material, which seems to be smooth dorsally. Cullison's material consists of internal moulds, and the pleural furrows are stronger than they are on the dorsal surface. In fact, the testate surface of our pygidium displays no defined furrows at all. Another incomplete pygidium is similarly smooth and shows the recurved doublure typical of this bathyurine genus.

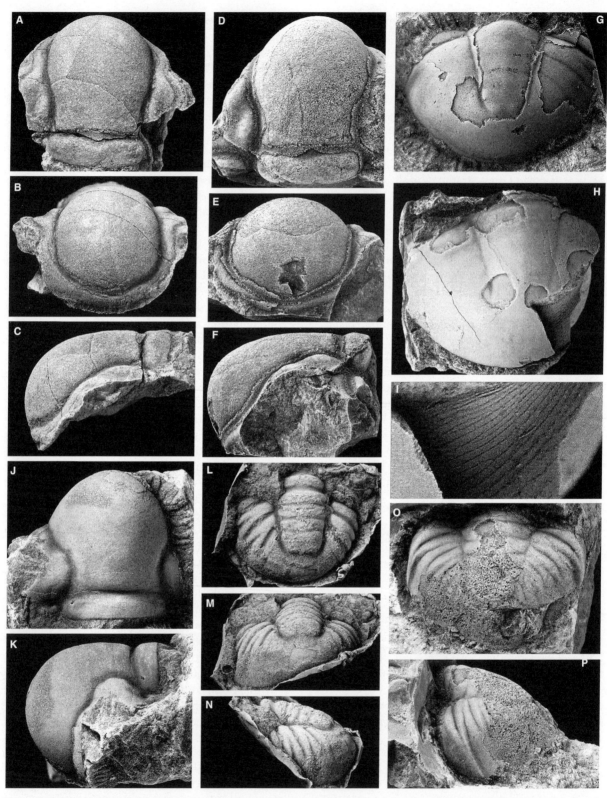

Fig. 8. **A–I**, *Bolbocephalus stclairi* Cullison, 1944. Upper Nordporten Member. **A–C**, Exfoliated cranidium CAMSM X 50188.42 in dorsal, anterior and lateral views (x2), *P. groenlandicus* fauna, ca. 105 m. **D–F**, Cranidium PMO 223.174 in dorsal, anterior and lateral views (x2), 105 m. **G**, Pygidium PMO 208.090 in dorsal view (x4), 150 m. **H**, Pygidium PMO 208.201 in dorsal view (x4), 190 m. **I**, Enlargement of H showing structure. **J–P**, *Bolbocephalus* sp. cf. *B. kindlei* Boyce, 1989. Lower part of Nordporten Member, *P. groenlandicus* fauna, H collection low in *P. groenlandicus* fauna, exact horizon unknown. **J–K**, Cranidium PMO 221.242 in dorsal and lateral views (x2.5). **L–N**, Latex cast of incomplete pygidium PMO 221.240 in dorsal, posterior and lateral views (x2). **O–P**, Pygidium PMO 221.241 in dorsal and lateral views (x2), poorly preserved posteriorly.

Bolbocephalus sp. cf. *B. kindlei* Boyce, 1989

Figure 8J–P

Material. – Pygidia, PMO 221.240, 208.241; possible cranidium, PMO 221.242.

Stratigraphic range. – H collection, low in Nordporten Member, presumed to be low *P. groenlandicus* fauna.

Discussion. – No formal description is possible on the basis of two large pygidia. However, it clearly differs from all others figured herein in the very deep expression of the lateral ends of the interpleural furrows, coupled with a markedly truncate termination of the pygidial axis. Both these features are shown on *Bolbocephalus kindlei* Boyce, 1989, which Boyce identified both from lower part of the Catoche Formation, western Newfoundland, and from the Cape Weber Formation, east central Greenland (Poulsen, 1937, pl. 8, fig. 2). Some of Boyce's specimens (e.g. Boyce 1989, pl. 25, fig. 5) display complete interpleural furrows, and the pleural furrows may be distinctively hooked backwards at their outer ends. Others, including the holotype (Boyce 1989, pl. 26, figs 1–4), have the interpleural furrows expressed only distally, at least on the anterior segments. The specimens from the Nordporten Member are like the holotype in this regard. However, the most posterior interpleural furrow extends all the way to the pygidial margin in all the specimens illustrated by Boyce (1989), whereas it fades out on our specimens. Because we are uncertain of the intraspecific variation that may apply in this taxon, our identification is qualified. A well-preserved cranidium from the same bed is possibly associated. Because the material is sparse and Boyce did not recognise the cranidium of *B. kindlei,* we are obliged to be cautious in this regard.

Genus *Catochia* Fortey, 1979

Type species. – *Catochia ornata* Fortey, 1979, Catoche Formation, western Newfoundland, original designation.

Discussion. – *Catochia* was described from two species collected from the St George Group, western Newfoundland, but has not hitherto been identified elsewhere. Two species are present in the Kirtonryggen Formation, although one is known from enough material to be named formally as new. The genus stands apart from other bathyurines with its small pygidium (which is hard to find in 'crack out' material) and slender genal spines; the type species has a surface sculpture on the glabella more usual on bathyurellines. Free cheeks, with their characteristic bevelled borders, are more easily collected than cranidia and differ critically from species to species. It is possible that a pygidium described tentatively as hystricurid by Poulsen (1937, pl. 8, fig. 3) could be that of a *Catochia* species.

Catochia hinlopensis n. sp.

Figure 9A–J

Holotype. – Free cheek, PMO 223.131 (Fig. 9E); Nordporten Member 190 m.

Material. – Cranidia, PMO 223.127, 208.208; free cheeks, PMO 221.232, 223.192; pygidium, PMO 208.204.

Stratigraphic range. – Upper part of the Nordporten Member 105–190 m, top *Petigurus groenlandicus* fauna to *Petigurus nero* fauna.

Diagnosis. – *Catochia* species with genal sculpture comprising a broad band of tubercles beneath the eye, lacking on rest of free cheek, and fine lines on outside of genal border. Glabella with rather widely scattered tubercles; broad palpebral rims. No occipital spine.

Etymology. – From the type locality along Hinlopen Strait.

Description. – Like other *Catochia* species, a small bathyurid, cephalon probably no longer than 7 mm. Glabella with sub-quadrate appearance, anterior lobe being somewhat transverse compared with many other bathyurids. Length (sag.) of glabella 1.5–1.65 times max. pre-occipital width, less transversely convex than *Jeffersonia* spp., frontal lobe curved down anteriorly to overhang border. Distinct and deep axial furrows bowed slightly outwards at the level of palpebral lobes. Occipital ring about 15–20% wider (tr.) than glabella in front of it. Glabella carries scattered, low tubercles, which are sparse along the mid-line. They are also absent over oval, backward-inclined areas on the flanks of the glabella opposite the posterior ends of the palpebral lobes, representing the posterior-most glabellar furrows; sub-circular impressions in front of them may be second

Fig. 9. **A–J**, *Catochia hinlopensis* n. sp. Nordporten Member, *P. nero* and *P. groenlandicus* faunas. **A–C**, Cranidium PMO 208.208 in dorsal, lateral and anterior views (x5), 190 m. **D**, Free cheek PMO 223.192 in lateral view (x5), 105 m. **E**, Holotype, free cheek PMO 223.131 in lateral view (x7), 190 m, showing characteristic surface sculpture. **F–G**, Cranidium PMO 223.127 in dorsal and lateral views (x7), 105 m. **H–I**, Free cheek PMO 221.232 in dorsal and lateral views (x9), 190 m. **J**, Incomplete pygidium PMO 208.204 in dorsal view (x6), 190 m (it was not possible to prepare this further). **K–O**, *Catochia ornata* Fortey, 1979. Nordporten Member *P. nero* fauna. **K–L**, Free cheek CAMSM X 50188.20 in dorsal and lateral views (x7) showing single line of tubercles, unspecified horizon. **M–N**, Latex cast of free cheek PMO 223.241 in lateral (x8) and dorsal (x10) views, horizon in upper part of Nordporten Member uncertain. **O**, Free cheek missing genal spine but showing border sculpture PMO 208.200 in dorsal view (x7), 190 m.

pair of furrows. Anterior part of occipital ring is smooth behind the deep occipital furrow, with tubercles concentrated on the lateral, gently inflated part of ring and along posterior margin. There is a prominent median tubercle but apparently no occipital spine. Large and deeply curved horizontal palpebral lobes, one-third (exsag.) glabellar length (sag.), with depressed rims with marginal raised selvage. Downward sloping pre-ocular cheeks very narrow (tr.) defined by sub-parallel sutures. Post-ocular sutural branches diverge at right angles to sag. line to define strap-like post-ocular cheek, which is at least half width (tr.) of occipital ring. The posterior fixed cheek is bisected by deep border furrow, but

neither the convex posterior border nor the genal field in front of it carry tubercles. Glabella overhangs pre-glabellar field reduced to almost nothing medially, and with exceedingly narrow convex border in front of it. Anterior of cephalon is arched about mid-line with genal borders on either side.

Free cheek is very distinctive, outline a quarter circle, with sharp, narrow genal spine at its lower edge. Genal field convex and divided into two parts: an upper part with tubercles and a smooth lower part. There is a narrow smooth band also beneath the eye, but exterior to this band lines of tubercles run approximately parallel to the base of the eye. The number of tubercles is variable: from two approximate rows to four rows. The sloping part of the genal field exterior to this band is again smooth, so there is an external unornamented band of variable width extending to the base of the genal spine. Lateral border narrows rapidly at its anterior end, becoming more convex in the process. The rest of the border is not particularly convex, bevelled towards the outside, defined by a wide border furrow which is deepest anteriorly and interrupted by the base of the genal spine. Posterior border is short (tr.), and its defining border furrow is deep, stopping before genal spine. The spine itself is flattened and rapier like, with the sharp edge facing laterally, gently tapering to sharp tip. The border carries a surface sculpture of very fine parallel lines on the outside of the lateral border, which run parallel to the cephalic edge. The base of the genal spine is swollen, originating quite high on the cheek, and carries much stronger raised lines running along its proximal length and at right angles to those on the border. Strongly curved, large eye lobe is high – at least as high as the lateral border viewed from the side.

Small pygidium known from one example and that not quite complete. With only two axial rings, the anterior of which carries a median spine, as in the type species. Long terminal piece. Flat, smooth border half length of axis. Gently convex pleural fields carrying two backward-sloping pleural furrows. Axial sculpture of fine tubercles. The proportions of the pygidium in this and the type species are comparable with the posterior part of pygidia of such bathyurids as *Jeffersonia striagena* n. sp., which like nearly all bathyurids has four pygidial segments. It does seem a possibility that in *Catochia*, two additional segments were 'released' into the thorax during later ontogeny.

Discussion. – We have chosen a free cheek as type specimen for this species, because of its several diagnostic features. The cranidium we have associated is a usual form for a bathyurine. Although we regard the association of sclerites as likely, we should note that the cranidium lacks an occipital spine, which is present on the type species and on *C. glabra* Fortey, 1979. On the other hand, the pygidium has a median anterior spine, like *C. ornata* Fortey, 1979. The tuberculate surface sculpture at once distinguishes *C. hinlopensis* from *C. ornata* and *C. glabra* from the Catoche Formation, western Newfoundland, the first with terrace ridges on the glabella, the second smooth and tending to effacement. All three have different free cheeks. There is some variation in the tuberculation on the cheek in *C. hinlopensis* with some specimens having a broader band than others. They all share the wide lateral genal border, and for

the moment, there is no reason to suspect more than one species, although the stratigraphic range is extended. However, a *Catochia* species with a single line of tubercles on the cheek and a narrow genal border does represent the second *Catochia* species in the upper part of the Nordporten Member, as described below.

Catochia ornata Fortey, 1979

Figures 9K–O

1979 *Catochia ornata* sp. nov. Fortey, pp. 81–84, pl. 27, figs 1–12.

1989 *Catochia ornata* Fortey, 1979; Boyce, p. 49, pl. 26, figs 5, 6.

Material. – Free cheeks, PMO 208.200, 223.241, CAMSM X 50188.20, X 50188.74.

Stratigraphic range. – Nordporten Member, 150–190 m *Petigurus nero* fauna.

Discussion. – The free cheek of *C. ornata* is unique in having a distinctive, single line of ornament obliquely following the base of the eye. In this character, it obviously differs from the tuberculate cheek of *C. hinlopensis* described above. We have only been able to find free cheeks of this species in the upper part of the Nordporten Member, which are figured here for comparison with the full description of Fortey (1979) and the illustration of a similar cheek from Cape Norman, western Newfoundland by Boyce (1989). This feature is considered sufficiently important to make a positive identification in the absence of other sclerites.

Genus *Jeffersonia* Poulsen, 1927

Type species. – *Jeffersonia exterminata* Poulsen, 1927, from the Nunatumi Formation, northwest Greenland.

Discussion. – The type species of *Jeffersonia* is a pygidium from strata of comparable age with the upper part of the Kirtonryggen Formation or the lower part of the Valhallfonna Formation. The problem of inadequate bathyurid type material has been discussed previously (p. 21), and this genus provides an example. Fortey (1979) placed *Bathyurus timon* Billings, 1865, from the Catoche Formation of western Newfoundland into *Bathyurina* Poulsen, 1937 (type species, *B. megalops* Poulsen, 1937, from East Greenland) because of the similarity of the appropriate sclerite of Billings' species to the type cranidium on which *Bathyurina* is based. However, the pygidium of *B. timon* is closely similar to that of *Jeffersonia exterminata*, being elongate with three well-defined convex pleural segments, a distinct border and lacking a terminal spine. In recognition of this, Fortey (1986) reassigned *Bathyurus timon* Billings to the senior genus, *Jeffersonia*. The assumption might well be that *Bathyurina* Poulsen, 1937, is a junior synonym of *Jeffersonia* Poulsen, 1927, but this can only be proved by discovery of more complete material of the type species of both genera. Knowledge of the cranidium of *Jeffersonia exterminata* is particularly important in this regard. For the moment, the pragmatic approach is to use the name with priority, *Jeffersonia*, to apply to species from Spitsbergen having pygidia similar to the type of that genus. This is not unreasonable, given the wide geographical distribution of species along the eastern margin of Ordovician Laurentia, and has been followed by other authors such as Boyce (1989) and Loch (2007). The association of sclerites of the species from Spitsbergen and Newfoundland is secure, whatever the ultimate nomenclatural situation. Several species from the Early Ordovician of Missouri named by Cullison (1944) have been neglected since their publication; these are reillustrated here for comparative purposes as relevant.

Jeffersonia cranidia are all very similar, but because they are thick shelled, internal moulds can appear quite different from testate surfaces. They have large palpebral lobes that are exceptionally strongly curved through more than a semicircle. The occipital ring has lateral lobes that increase the transverse width compared with the glabella immediately in front, but these are often hard to see on internal moulds. Cranidia differ from species to species in the development of the pre-glabellar field, and in details of the anterior border, which is thin and often very hard to prepare. Free cheeks show interesting characters on the border, and genal spines can be variably developed. Pygidia usually show three deep axial rings on specimens with cuticle, and a fourth is clear on internal surfaces, together with a prominent semicircular terminal piece. Overall pygidial outline, width of the border and its convexity, and the development of pleural lobes, varies from one species to another. It is also notable that *Jeffersonia* can attain quite a large size (cranidia to at least 3 cm) when the glabella tends to be more acuminate at the front. Here we exclude *J. granosa* Cullison, 1944, from *Jeffersonia*, unlike Loch (2007), as discussed further below.

Jeffersonia striagena n. sp.

Figures 10, 11

Holotype. – Pygidium, PMO 208.140 (Fig. 11G); 105 m from base of Nordporten Member.

Material. – Cranidia, PMO 223.140, 223.146, 223.236, 223.147, 223.188, CAMSM X 50188.57; free cheeks, PMO 208.126–8, 223.169, 223.178, 223.187b, 223.220, PMO NF 1913; hypostomes, PMO 223.149, 223.180, 223.214(2), 223.234(2); pygidia, PMO 208.096, 208.121, 208.212, 208.214, 223.142–4, 223.148 PMO NF 1911, CAMSM X 50188.75.

Stratigraphic range. – Lower part of the Nordporten Member, 70 m horizon to 130 m from base associated with *Petigurus groenlandicus* fauna.

Diagnosis. – *Jeffersonia* species with relatively subdued surface sculpture of flattened tubercles weakly reflected on internal moulds. Pre-glabellar field short or absent. Inflated genal angle with prominent raised ridges; short, stout genal spine. Elongate pygidium almost pointed posteriorly, feeble sculpture on pleural fields and relatively wide, smooth border.

Etymology. – Latin *striagena*: 'striated cheek' – referring to the diagnostic swollen area on the free cheek.

Description. – The species is one of the commoner bathyurids in the lower Nordporten Member, and the sclerites are associated with confidence and type series from a single bed. Most specimens fall in the small-to-medium-size range with cranidia and pygidia both between 5 mm and 1.5 cm long, but there is evidence of continued growth, with free cheeks up to 3.5 cm long, suggesting that a fully mature trilobite might be ca. 10 cm long. All material is prone to exfoliate – the cuticle is not particularly thick – and the internal mould reflects the dorsal surface features with variable fidelity. Because the specimens from the type horizon show a range of subtle variation, a large number of specimens are illustrated.

Cranidium convex, with the anterior part of the glabella having a bowed and hardly, to slightly recurved profile at front, fixed cheeks sloping steeply forwards in front of the eyes and posterior part of cheeks less so. Forward curvature of glabella may conceal anterior cranidial border in dorsal view. In this view, maximum transverse width level with palpebral lobes is between 60 and 70% sagittal length, longer glabellas being on larger specimens. Narrow

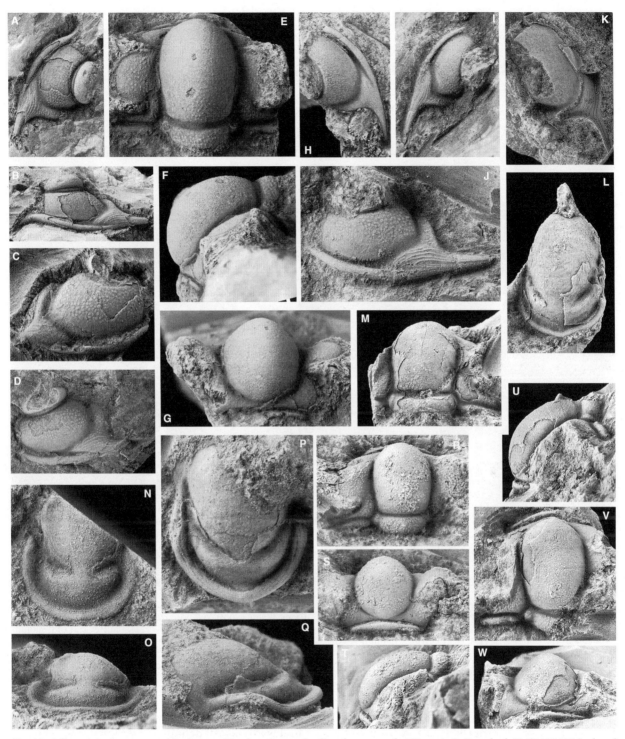

Fig. 10. Jeffersonia striagena n. sp. Nordporten Member, *P. groenlandicus* fauna, mostly 105 m. **A–B**, Free cheek PMO 223.178 in dorsal and anterolateral views (x2). **C**, Free cheek PMO 208.128 in lateral view (x2). **D**, Free cheek PMO 208.126 in lateral view (x3). **E–G**, Cranidium PMO 223.236 in dorsal, lateral and anterior views (x7). **H**, Internal mould of free cheek PMO 223.187b in dorsal view (x3) showing poor reflection of dorsal sculpture on internal surface. **I–J**, Free cheek, PMO 223.220 in dorsal (x4) and lateral (x6) views, to show ridges on inflated border. **K**, Free cheek PMO 208.127 in dorsal (x3), 70 m, view with doublure prepared. **L**, Hypostome PMO 223.149 in ventral view (x5). **M**, Cranidium PMO 223.146 in dorsal view (x3). **P–Q**, Hypostome PMO 223.180 in ventral and lateral views (x6). **R–T**, Small cranidium CAMSM X 50188.57 in dorsal, anterior and lateral views (x5), *P. groenlandicus* fauna, exact horizon unspecified. **U–W**, Cranidium with finer tubercles PMO 223.140 in lateral, dorsal and anterior views (x4). **N–O**, Incomplete hypostome attributed to *Uromystrum affine* PMO 223.224 in dorsal and posterior views (x10), for comparison.

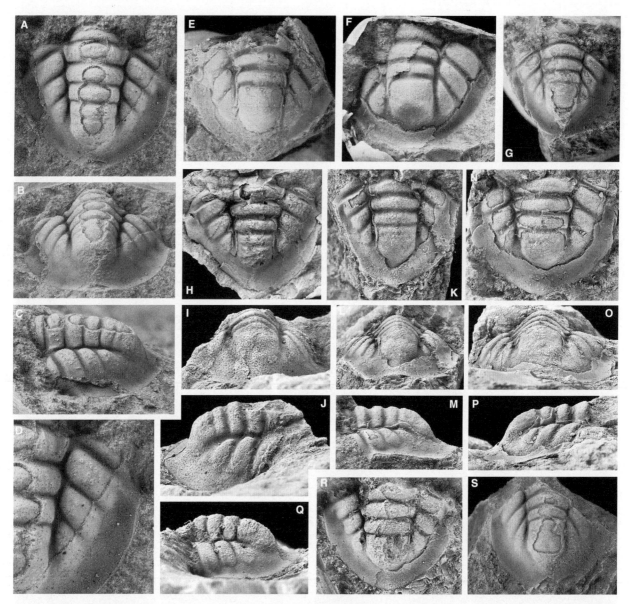

Fig. 11. Jeffersonia striagena n. sp. Nordporten Member, *P. groenlandicus* fauna, 105 m., pygidia to illustrate variation. **A–C**, Pygidium CAMSM X 50188.75 in dorsal, posterior and lateral views, (x4), ca. 105 m; **B** shows postaxial tubercles. **D**, Enlargement of **A** showing sculpture of weak pleural tubercles and pitted internal mould. **E**, Pygidium PMO 208.212 in dorsal view (x4). **F**, Latex cast of pygidium PMO 208.214 in dorsal view (x3). **G**, Holotype, pygidium PMO 208.140 in dorsal view (x3). **H**, Pygidium PMO NF 1911 in dorsal view (x3), exact horizon unknown. **I–J**, Pygidium PMO 223.142 in posterior and lateral view (x5). **K–M**, Pygidium PMO 223.143 in dorsal, posterior and lateral views (x5). **L** shows median inflection of border. **N–P**, Pygidium PMO 223.144 in dorsal, posterior and lateral views (x3), slightly wider specimen with less sloping pleural furrows. **Q–R**, Pygidium PMO 223.148 in lateral and dorsal views (x5). **S**, Pygidium PMO 208.121 in dorsal view (x2.5)

and low point on the glabella is just in front of the occipital ring, forward from which the axial furrows are bowed outwards and transverse glabellar convexity increases. Axial furrows are moderately well defined and shallower around frontal glabellar lobe, which comes to a point on the mid-line. Glabellar furrows not defined, though there is a smooth area opposite the palpebral lobe on the flanks of the glabella that probably represents an area of muscle insertions. The occipital furrow is deep, slightly

backward bowed, defining an occipital ring, which is 20% or somewhat less dorsal length of glabella (including occipital ring). The most prominent feature of the occipital ring is the presence of lateral lobes, which bulge out laterally causing a sharp deflection in the posterior course of the axial furrows, and defined by weak furrows on their inner sides. The lobes appear to be weaker on small cranidia. The adjacent area of the posterior part of the fixed cheek is also inflated – an area narrowing forwards on to the palpebral region. Palpebral lobes large, half cranidial length (exsag.), flat, sloping inwards and highly curved, making more than a semicircle, with narrow, uptilted palpebral rims running along their whole length. Pre-glabellar field very short on sagittal line or can be reduced to nothing. Facial suture curves outwards and diverges at a high angle (70°) to the sagittal line behind palpebral lobe and at a relatively low angle in front (20–30° viewed dorsally) in front of it. In anterior view, it can be seen to curve rather evenly adaxially adjacent to the downsloping part of the anterior fixed cheek to cut the anterior border at an obtuse angle. Anterior border furrow is very deep and narrow, and convex, narrow, ledge-like border returns abruptly from it, having a short, anterior facing outer edge, and overall gently arched across mid-line. Post-ocular cheek acutely triangular, not as wide (tr.) as occipital ring, with deep posterior border furrow separating off narrow and convex posterior border which widens away from axial furrow.

Free cheek viewed dorsally with outline approximating a quarter circle leading into stout genal spine. Deep lateral border furrow where abutting fixed cheek but becoming wider and somewhat shallower distally and then abruptly narrower and fading into proximal part of genal spine. It is likely that the true border furrow branches off the lateral border furrow and passes across the genal corner as a broad, relatively shallow furrow to join the posterior border furrow. This is by analogy with other bathyurid cheeks. The lateral border is convex with a steep outer slope. Posterior border less convex and posterior border furrow distinct, but wider than the lateral border furrow at its narrowest. Genal field is convex, somewhat bulging. The area between the abaxial end of the posterior border furrow and the margin is swollen into a triangular, tumid region, which, on its dorsal surface, carries a series of six to eight remarkably wide, raised ridges running back sub-parallel to the lateral genal margin. Stout genal spine often slightly inturned towards its tip, with steeply oval cross section, and no longer than eye. The genal corner swelling is visible also on internal moulds. Since the doublure, which carries a few terrace ridges, is reflexed dorsally, it forms a tube running along the anterior cephalic margin. However, its outline does not follow that of the dorsal inflated area, so this species had a distinctive ventral chamber on the posterolateral edge of the free cheek. Eye has a high visual surface (cf. many bathyurellines) and a prominent convex narrow eye socle, narrowing anteriorly, which is isolated from the rest of the cheek by a deep furrow. Tiny lenses were presumably orientated with a largely lateral field of view.

Hypostome associated with moderate confidence, two-thirds as wide as long, with a prominent middle body divided into a long oval anterior lobe, and a short lunate posterior lobe. Middle furrows long, converging inwards at about 45° to sag. line. Narrow posterior border convex and rounded much like cranidial anterior border. Anterior wings relatively short.

Pygidium 1.1–1.2 times wider than long with convex flanks sloping down to finally flattened border. Although no specimens have a posterior spine, there is a rounded median acumination that can lend some specimens a sub-triangular outline. The well-defined gently tapering axis extends to three-quarters pygidial length and is evenly convex (tr.), but almost merges with post-axial field posteriorly. Three rings are invariably clearly defined, but the terminal piece beyond shows a faint impression of a fourth in some specimens. Terminal piece accounts for 40–45% axial length. Three straight, deep, narrow pleural furrows, with the suggestion of a fourth, stopping at border. There is variation in the angle at which these furrows slope backwards, some markedly more than others (Fig. 11S). The latter have relatively wider borders. This is regarded as variation within the species. Pleural bands are terminated at the border sharply in some specimens, others more gently merging. Relatively wide border is dished, with the outer edge nearly horizontal.

Surface sculpture is not strongly developed and again somewhat variable. The degree to which it is reflected on the internal mould also seems to vary. Some cranidia are nearly smooth; others display fine tuberculation on the glabella. Low tubercles on the post-ocular cheeks and scattered on palpebral lobes. Free cheeks are characteristically tuberculate outside the furrow defining the eye lobe, but with a smooth band along the periphery of the genal field. The border has fine parallel raised lines, but the inflated area posterolaterally carries strikingly prominent raised lines. The pygidial border is invariably smooth, while the axis and pleurae may have low tubercles that are often not reflected on internal moulds.

Discussion. – The inflated posterolateral areas on the free cheeks provide a good specific character in comparison with other species assigned to *Jeffersonia* for which free cheeks are known. The feature can, however, be matched on the early Whiterockian (Dapingian) genus *Psephosthenaspis* (see Fortey & Droser 1996; Adrain *et al.* 2012). This genus includes four species with transverse pygidia and forwardly expanding, anteriorly truncate glabellas, and Adrain *et al.* (2012) describe the genus as a bathyurid clade not yet known before the Middle Ordovician. In this case, the genal structure on *J. striagena* would likely be homoplastic. However, it is also worth considering that sister species of *Psephosthenaspis* are to be found among taxa that might otherwise be assigned to *Jeffersonia* on the majority of their features; more evidence is needed from later Floian (Blackhillsian) strata to identify possible connecting species.

As far as the pygidium is concerned, *J. striagena* differs from the type species in its more well-defined, wider pygidial border, tending to a triangular outline, and more subdued axial sculpture. Subdued sculpture, more triangular outline with wider border, and a generally better-defined fourth axial ring distinguish the pygidium of *J. striagena* from that of *J. timon*. The older species *J. viator* n. sp., described below, has a longer pre-glabellar field and a much more transverse pygidium with a wider, concave border, as well as a different border of the free cheek, with an inflated area extending further forwards. The most similar of several other named *Jeffersonia* species is *J. angustimarginata* Boyce, 1989, from the lower part of the Catoche Formation, western Newfoundland, the holotype pygidium of which does show a weak terminal acumination, and subdued surface sculpture. The material of this species is very limited, and the free cheek is incomplete so that the diagnostic border and genal spine of *J. striagena* cannot be seen. The latter is more tuberculate than that of *J. striagena*, and it is possible to make out tubercles on the inner part of the lateral border (Boyce 1989, pl. 27, fig. 5), which is unlike *J. striagena*. A second pygidium attributed to *J. angustimarginata* by Boyce is both coarsely tuberculate and narrower than the holotype and seems likely to belong to another species. In view of these ambiguities, and because we have much better-preserved and abundant material associated from a single horizon, we propose a new species herein, cognisant that its status may be revised when more material is collected from western Newfoundland.

Jeffersonia viator n. sp.

Figure 12A–B, D–K, N–S, X–Y

?1944 *Jeffersonia* sp. 1 Cullison 1944, p. 77, pl. 35, figs 9, 10.

Holotype. – Pygidium, PMO 223.201 (Fig. 12N–O); uppermost Bassisletta Member, *Chapmanopyge* fauna.

Material. – Cranidia, PMO 223.202, 223.216, PMO 221.259, CAMSM X 50188.76; free cheeks, PMO 223.218, 223.221; pygidia, PMO 223.201, 223.207–8, CAMSM X 50188.77.

Stratigraphic range. – Upper 34 m of Bassisletta Member, *Chapmanopyge* fauna, and basal part, possibly up to 50 m of Nordporten Member, *Petigurus groenlandicus* fauna.

Diagnosis. – *Jeffersonia* with short, but distinct pre-glabellar field; librigenal border with posterolateral, inflated triangular area inside border; pygidium transverse, up to 1.5 times wider than long with comparatively wide, smooth border, posterior pleural furrows reduced or absent; surface sculpture finely tuberculate, especially axially.

Etymology. – Latin: 'a traveller' – referring to its possible wide occurrence.

Description. – The species is closely similar to *J. striagena*, which it underlies stratigraphically. Differential details are as follows:

1   There is a distinct pre-glabellar field across the mid-line, which is vertical, or even recurved beneath the glabella. On *J. striagena*, the pre-glabellar field is much reduced or absent.

Fig. 12. **A–B, D–K, N–S, X–Y,** *Jeffersonia viator* n. sp. All top 2 m of Bassisletta Member, *Chapmanopyge* fauna, except **P**, which is 34 m below the top of the Bassisletta Member, and **I–J** and **Y** that are low Nordporten Member, *P. groenlandicus* fauna ca. 50 m. **A–B,** Free cheek PMO 223.221 in dorsal and lateral views (x4), **D,** Latex cast of cranidium to show palpebral lobe PMO 223.216 in dorsal view (x7). **E, K,** Free cheek PMO 223.218 in dorsal view (x5) and in lateral view to show sculpture (x6). **F–H,** Cranidium PMO 223.202 in dorsal, anterior and lateral views (x4). **I–J,** Cranidium CAMSM X 50188.76 in dorsal and anterior views (x6), low Nordporten Member ca. 50 m. **N–O,** Holotype, pygidium PMO 223.201 in dorsal (**N** x6) and posterior views (**O** x4). **P,** Latex cast from large fragmentary cranidium PMO 221.259 in dorsal view (x1.5), Nordporten Member, 60–70 m. **Q–S,** Incomplete pygidium PMO 223.207 in dorsal, posterior and lateral views (x3). **X,** Cast of incomplete pygidium PMO 223.208 in dorsal view (x2). **Y,** Pygidium CAMSM X 50188.77 in dorsal view (x4), low Nordporten Member ca. 50 m. **C, L–M,** *Jeffersonia crassimarginata* Cullison, 1944. **C,** Free cheek, original of Cullison, 1944, pl. 35 fig. 14, YPM 17404 in lateral view (x3). **L–M,** Pygidium, holotype of Cullison, 1944, pl. 35 figs 15–16, YPM 17402 in dorsal and posterior views (x4), internal mould. Note: the cranidium was too poorly preserved to merit refiguring. **T–V,** *Jeffersonia* sp. 1 (Cullison, 1944), internal mould of pygidium, original of Cullison, 1944, pl. 35 figs 9–10, USNM 155402 in dorsal, posterior and lateral views (x3). **W,** *Jeffersonia* sp. 2 (Cullison, 1944), pygidium, original of Cullison, 1944, pl. 35 figs 7–8, USNM 155403 in dorsal view (x2).

2 The free cheek has a similar inflated postero-lateral triangular area in front of the genal spine, but it extends further forwards, petering out within the lateral border furrow opposite the forward end of the eye. It does have a sculpture of ridges, but these are finer and

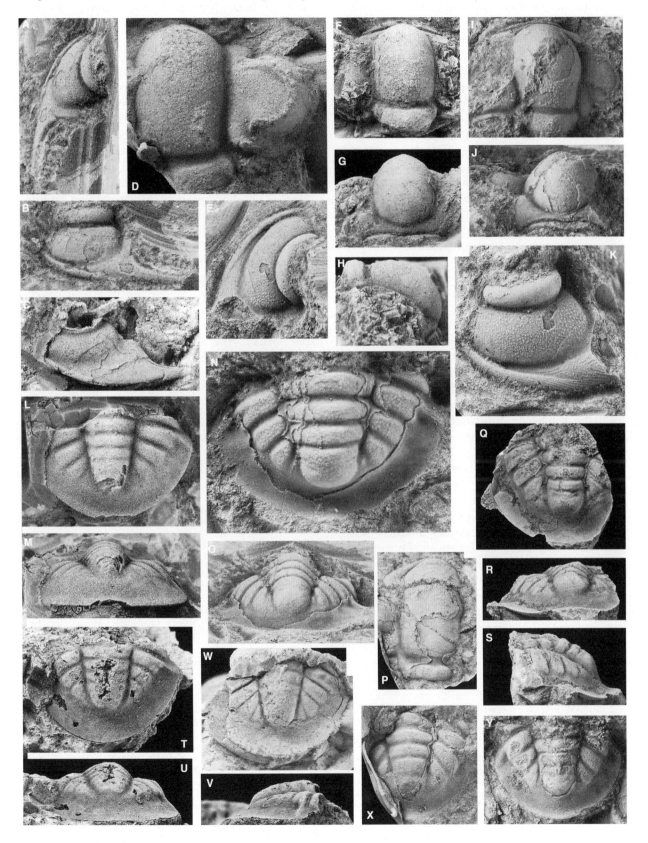

more numerous than is the case in *J. striagena*.

3 Pygidia include examples, such as the holotype, which are two-thirds as long as wide and wider than any specimen of *J. striagena*. Larger specimens can be relatively longer, but are still wider than those of *J. striagena*.

4 Pygidial border is concave posteriorly, wider than that of *J. striagena*, especially laterally, and does not come to a posteromedian point.

5 Surface sculpture consists of generally distributed dense, course granulation, not extending on to posterior pygidial border.

There is some variation in the expression of the fourth pygidial pleural furrow, which can be very short on some specimens, outlining a diminutive posterior pleural lobe. Internal moulds may show the pygidial ribs bisected longitudinally by weak median furrows.

Discussion. – Although we do not have a lot of material of this species, it appears to be a distinctive member of the genus of stratigraphic significance and hence merits formal recognition. Its wide, concave pygidial border sets it apart from other named species. It is closest to *J. striagena* of the Spitsbergen species, with which it has been compared above. *Jeffersonia ulrichi* Loch, 2007, from the Kindblade Formation, Oklahoma, has a border as wide, but it slopes steeply down to the margin, and this species also has a less forwardly protuberant glabella. The types of *J. crassimarginata* Cullison, 1944, from the Theodosia Formation (Blackjack Knob Member) of Missouri are illustrated for comparison (Fig. 12C, L–M), since the pygidia of this species display a wide border. However, the pygidial border essentially continues the downward slope of the proximal part of the pleural field, which is more typical of *Peltabellia* with which it has been compared above. The free cheek of this species displays a wide, triangular area in front of the genal spine, defined by the lateral border furrow at the edge of the genal

field. This is likely to represent the plesiomorphic condition compared with *Jeffersonia striagena* and *J. viator*, in which this area is divided into an inflated inner portion and an exterior 'pseudo-border' that continues near the margin towards the genal spine. Pygidia of two species figured under open nomenclature by Cullison (1944) as *Jeffersonia* sp. 1 and sp. 2 offer a better comparison as far as width of pygidial border is concerned and are refigured here on Figure 12T–W. *Jeffersonia* sp. 2 has narrow pleural furrows, and the border is not clearly demarcated from the pleural lobes; as with *J. crassimarginata*, it might be better assigned to *Peltabellia*. However, *Jeffersonia* sp. 1 is like a larger example of *J. viator* (Fig. 12Y), differing only in having the concave border somewhat more downsloping and less firmly impressed furrows. The latter difference is probably attributable to the internal mould preservation, and it is considered possible that the same species is present in Missouri in the Theodosia Formation as in the top of the Bassisletta Member in Svalbard, with the usual caveat that we need to know more of cephalic sclerites of the former. We also note the possible identity with a species described as *Jeffersonia* sp. from Ellesmere Island by Adrain & Westrop (2005, fig. 10.14–10.16, but described as 'Fig. 10N–P' in text), especially with regard to the pygidium. The boundary between the more or less plesiomorphic species assigned to *Peltabellia* and the derived species that can be placed in *Jeffersonia* requires further critical analysis.

Jeffersonia timon (Billings, 1865)

Figure 13A–M

1865 *Bathyurus timon* (n. sp.) Billings, p. 261, fig. 244.

1986 *Jeffersonia timon* (Billings, 1865); Fortey, p. 18, pl. 1, fig. 6.

Fig. 13. **A–M**, *Jeffersonia timon* (Billings, 1865), Upper part of Nordporten Member, Upper *P. groenlandicus* – *P. nero* fauna. **A**, Cranidium PMO NF 1895 in dorsal view (x7), exact horizon unknown. **B–C**, Pygidium PMO 223.176 in dorsal and posterior views (x4), 105 m. **D–F**, Cranidium PMO 223.179 in dorsal, anterior and lateral views (x5), 105 m. **G–I**, Pygidium CAMSM X 50188.78 in dorsal, posterior and lateral views (x4), *P. nero* fauna, unspecified horizon. **J–L**, Pygidium CAMSM X 50188.79 in dorsal, posterior and lateral views (x3), unspecified horizon. **M**, Pygidium PMO 208.225 in dorsal view (x1.5), 130 m. **N–S**, *Jeffersonia jennii* Cullison, 1944. **N–P**, Internal mould of pygidium YPM 17406 in dorsal, posterior and lateral views (x3). **Q–S**, Internal mould of cranidium YPM 17405 in dorsal, anterior and lateral views (x2). **T–V**, '*Jeffersonia*' aff. *J. granosa* Cullison, 1944. Cranidium PMO 221.266 in dorsal, anterior and lateral views, 130 m from Base of Nordporten Member, top *P. groenlandicus* fauna. **W–BB**, *Jeffersonia granosa* Cullison, 1944. Paratypes of Cullison. **W–Y**, Internal mould of cranidium YPM 17171 in dorsal, anterior and lateral views (x3). **Z–BB**, Imperfectly preserved pygidium YPM 17172 in dorsal, posterior and lateral views (x3).

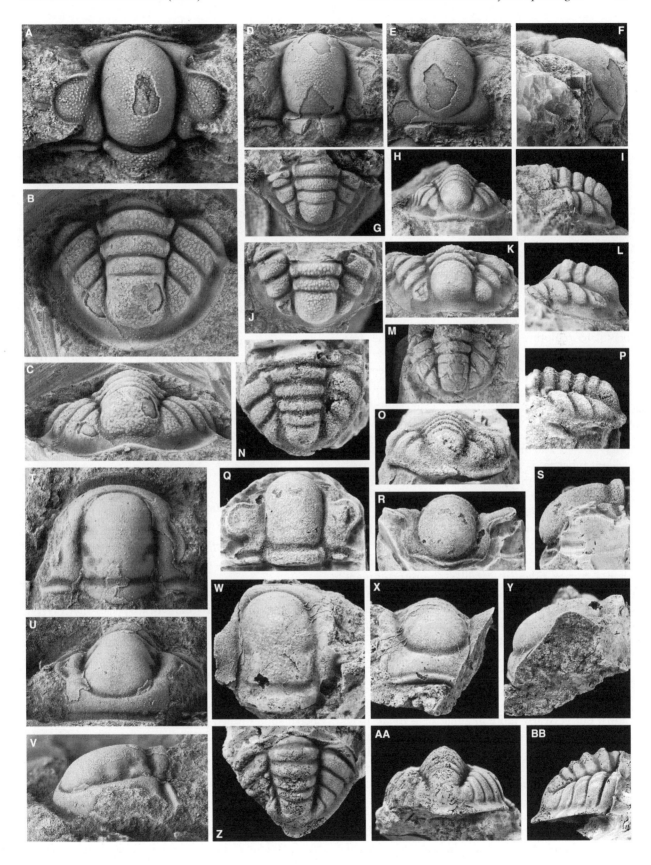

1989 *Jeffersonia timon* (Billings, 1865); Boyce, p. 51 (*cum syn.*)

Material. – Cranidia, PMO 223.179 NF 1895; pygidia, PMO 208.225, 223.176, CAMSM X 50188.78, X 50188.79.

Stratigraphic range. – Upper part of the Nordporten Member, ?105 m, 120–200 m, typically *Petigurus nero* fauna.

Discussion. – A full description by Fortey (1979) does not require repetition herein, particularly as most features are shared with *J. striagena* given previously. Like that species, the pre-glabellar field is reduced, but the pygidial border is relatively narrow along its length, and the fourth pair of pygidial pleural furrows is not discernible except possibly as a posterior widening of the pygidial axial furrows. Tuberculate sculpture is relatively coarse. The free cheek associated by Fortey (1979, pl. 25, fig. 9) does not show the posterolateral inflated area seen on *J. striagena*. Loch (2007) noted the similarity between this species and *J. jennii* Cullison, 1944, from the Cotter Formation of Missouri. The type specimens of this species are reillustrated here (Fig. 13N–R) and are internal moulds, so the lack of evident tubercles is not significant, since with a species with such a thick cuticle, the internal surface fails to reflect the sculpture. On this evidence, the cranidia of *J. jennii* and *J. timon* may be considered to be the same. Loch (2007, pl. 8, fig. 8) also illustrated lack of pre-glabellar field in *J. jennii*. However, the pygidial border of *J. jennii* is little more than a convex ridge (it was likely more prominent on material with cuticle) half the width of that of most internal moulds of *J. timon*. The fourth axial ring is more prominent on the internal pygidial mould of *J. jennii*, and the fourth furrow is just distinguishable from the axial furrow it abuts.

A well-preserved pygidium (Fig. 13B) from 105 m from the base of the Nordporten Member is from the upper part of the *Petigurus groenlandicus* fauna, but is clearly more like that of *J. timon* than *J. striagena*. It differs subtly from the type (Fortey 1979, pl. 25, fig. 3) in having a wider, sub-quadrate terminal piece, and a more upturned posterior margin, but the taxonomic significance of these differences is as yet uncertain. This specimen shows the smooth muscle pads on the anterolateral parts of the axial rings particularly clearly, which also explains the positioning of the lateral lobes on the occipital ring. For the moment, it is included as *J.* cf. *timon* and a likely cranidium associated.

'Jeffersonia' aff. *J. granosa* Cullison, 1944

Figure 13T–V

Material. – Cranidium, PMO 221.226.

Stratigraphic range. – Nordporten Member, 130 m from base, top of *Petigurus groenlandicus* fauna.

Discussion. – A single, well-preserved cranidium is not a good basis for formal taxonomy. It does, however, show short, deeply incised lateral glabellar furrows, together with an anteriorly truncate glabella, and a relatively long (tr.) concave anterior cranidial border. *Jeffersonia* species described above are united in *not* having incised glabellar furrows, and in their convex, deeply defined, cranidial borders. *J. granosa* Cullison, 1944, was described from cranidia and a pygidium from Missouri (Cullison's Rich Fountain Formation, reillustrated here Fig. 13W–BB), and further material was described by Loch (2007) from the Kindblade Formation, Oklahoma. This material shares several cranidial features with our specimen, but the Spitsbergen form has much less curved palpebral lobes, less emphatic sculpture, and a concave border, so is very unlikely to be conspecific. It cannot be named formally on the basis of our sparse material.

Genus *Peltabellia* Whittington, 1953

Type species. – *Jeffersonia peltabella* Ross, 1951, original designation.

Remarks. – We retain *Peltabellia* for a collection of stratigraphically early bathyurid species that may at the moment constitute a paraphyletic group, since they are united by generally primitive characters. Typical bathyurellines have flat or even concave pygidial borders; on *Peltabellia*, the border typically slopes downwards. On the other hand, some species that have been placed in *Peltabellia* have long, falcate genal spines much like those of *Licnocephala*, *Punka* and *Bathyurellus*. A rather sub-quadrate glabella seems to be typical of early bathyurids in general. As constituted at the moment, *Peltabellia* is therefore something of a taxon of convenience. However, a species from Spitsbergen is assuredly closer to previously named species of *Peltabellia* (notably the type species) than to any other bathyurid, and if *Peltabellia* is redefined in a more restricted sense, it seems likely that *P. glabra* n. sp. will prove a part of it. Loch (2007) proposed several new species that he assigned to *Strigigenalis* Whittington and Ross, in Whittington, 1953, which lacked the posteromedian pygidial spine that was typical of later species of that genus (Fortey 1979; Boyce (1989). The type species, *S. cassinensis* Whittington, 1953, also lacks such a spine (Brett & Westrop 1996, p. 423). If correctly interpreted as a single species, *Strigigenalis crassimarginata* (Cullison, 1944) evidently shows variable development of the posterior spine within several populations, although it is never much more than a stub (Loch 2007, pl. 16). There appears to be a stratigraphic sequence of some *Strigigenalis* spp. through the later Ibexian typified by a progressively lengthening pygidial spine, *S. crassimarginata* (Cullison 1944) – *S. brevicaudata* Boyce, 1989 – *S. caudata* (Billings, 1865); this sequence occurs along the length of the eastern Laurentian platform. This spe-

cies group seems to constitute a clade, and the broad-based posterior spine provides the easiest way to define it, but since the genus as a whole also includes species without spines, for some earlier taxa, the distinction from *Peltabellia* is not obvious.

Peltabellia glabra n. sp.

Figure 14

Holotype. – Pygidium, PMO 223.135 (Fig. 14H–J); Bassisletta Member.

Material. – Cranidia, PMO 223.136–7; free cheek, PMO 223.134; pygidia, PMO 223.138–9, PMO 223.239.

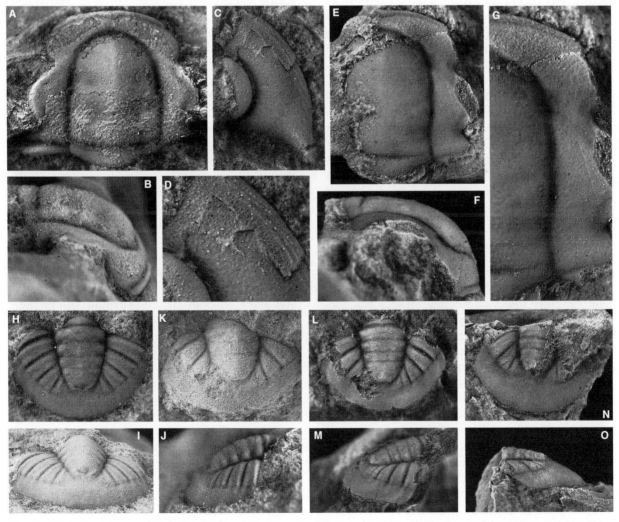

Fig. 14. Peltabellia glabra n. sp. Bassisletta Member, 60 m from base. *P. glabra* fauna. **A–B**, Cranidium PMO 223.136 in dorsal and lateral views (x12). **C**, Free cheek PMO 223.134 in dorsal view (x12). **D**, Enlargement of **C** showing the structure. **E–F**, Cranidium PMO 223.137 in dorsal and lateral views (x8). **G**, Enlargement of **E** showing the structure. **H–J**, Holotype pygidium PMO 223.135 in dorsal, posterior and lateral views (x8). **K**, Pygidium PMO 223.239 in dorsal view (x8). **L–M**, Pygidium PMO 223.138 in dorsal and lateral views (x6). **N–O**, Pygidium PMO 223.139 in dorsal and lateral views (x4).

Stratigraphic range. – Bassisletta Member of the Kirtonryggen Formation, 60 m from base, *Peltabellia glabra* fauna (early Tulean).

Etymology. – Latin: 'smooth' – referring to lack of sculpture.

Diagnosis. – *Peltabellia* of low convexity and lacking surface sculpture; palpebral lobes one-third cranidial length; genal spines short; four pairs of pygidial pleural furrows that stop at fairly narrow sloping border, which widens medially.

Description. – This is the only trilobite we have obtained from the lower part of the Bassisletta Member of the Kirtonryggen Formation where it occurs in association with algal laminates and stromatolitic structures. Most of the specimens are quite small, but there is no evidence to suggest that the species grew large. Preparation of the material is difficult as the rock is partly silicified, and sclerites tend to 'disappear' into the matrix. Transverse convexity of glabella is low, and pre-glabellar field slopes downwards less steeply than in other species of the genus. In dorsal view, glabella occupies 85% of cranidial length and is two-thirds as wide as long, more or less parallel sided, with rounded front, although slight median acumination visible in anterior view. Glabellar furrows are not defined, although smaller specimens show broad, sub-circular muscle scars opposite posterior part of palpebral lobe, and another pair opposite anterior end. Axial furrow narrow and of even width around perimeter of glabella. Occipital ring well defined, or fainter laterally, 12–17% total glabellar length, widest medially. Faint median furrow bisects pre-glabellar field. Palpebral lobes quite far removed from glabella, and faint eye ridge indicated running inwards and forwards to anterolateral corner of glabella. Palpebral lobes one-third cranidial length in dorsal view, and somewhat longer on small cranidia, moderately strongly curved and with rather ill-defined, slightly elevated rims. Facial sutures diverge weakly in front of and strongly behind the eyes, anterior sections curving rather evenly inwards and more strongly as they approach the anterior border. Post-ocular cheek at least two-thirds width of occipital ring, more or less bisected by border furrow. Anterior border narrow and convex, border furrow slightly inward curved on mid-line. Faint indications of caeca on anterior parts of fixed cheeks in front of eye ridges, otherwise no indication of surface sculpture.

Free cheek is narrow (tr.) for a bathyurid, almost a quarter circle in outline. Curved eye lobe is underlain by a narrow, convex eye socle. Border is hardly defined; it carries a few conspicuous raised ridges that are slightly oblique to the margin. Genal spine is short, stout and triangular, somewhat carinate on top. Gently convex genal field has weak caeca on parietal surface.

Pygidium 1.4–1.5 times wider than long in dorsal view, with axis occupying about 70% length and 40% width. Axis is gently convex sided and slowly tapering to transverse terminal piece. Half-ring prominent, half as wide as ring behind. Four axial rings of which the first three are visible across the axis, while the fourth is weakly indicated laterally, only slightly decreasing in length (sag.) posteriorly. Lateral parts of anterior rings slightly tumid. Four pygidial ribs of decreasing length posteriorly, delineated by strong, straight furrows. A weak fifth pair of furrows may be present, directed nearly exsagitally. These furrows stop at border; faint interpleural furrows on some specimens. Border slopes downwards, clearly demarcated from pleural field, and widens slightly behind the axis. An incomplete larger pygidium (Fig. 14N) shows a slightly wider border. No surface sculpture seen.

Discussion. – The type species *P. peltabella* (Ross, 1951) is from the 'Zone G' fauna of the Garden City Formation and is very much like the new species from Spitsbergen, although differing clearly in being more convex, having tuberculate surface sculpture, a prominent cranidial border and much narrower, less furrowed pygidial pleural fields. It shares with *P. glabra* the characters of short genal spines, rounded glabella front and sloping pygidial border, one which is clearly demarcated from the pleural fields. *Peltabellia elegans* Fortey & Peel 1990, from western North Greenland is also generally similar to *P. glabra*, but has a more rectangular glabella and steep pre-glabellar field, longer genal spines, larger eyes and a generally effaced pygidium with no more than three pleural ribs. Boyce (1989) and Adrain & Westrop (2005) have cautiously assigned to *Peltabellia* species with tumid pygidial ribs and flat pygidial borders that, while clearly related to one another, are much less like the three species cited above. Fortey & Peel (1990) noted that the Siberian genus *Biolgina* might be a junior synonym of *Peltabellia* and discussed several Russian species that are all more similar to *P. elegans* than they are to *P. glabra*. The new species is closely similar to *P. willistoni* Lochman (1966, pl. 62, figs 10, 11) from borehole material in the Williston Basin, Montana, attributed to 'Zone G'. Unfortunately, *P. willistoni* Lochman is known only from pygidia, but these do show strongly furrowed pleural fields similar to those of pygidia of *P. glabra*. However, comparing our material with the poorly illustrated material of Lochman (1966), it is quite secure that there are only three ribs on the Texan material and that the pygidial border is relatively wider. Lochman herself considered pygidia figured by Cullison (1944, pl. 34, figs 14, 15; pl. 35, figs 9, 10) under the name *Jeffersonia missouriensis* Cullison to be conspecific with her material. To add further complication, the nomenclature of this species was examined by Ross (1951, p. 76) who concluded that Cullison had attributed the wrong pygidium to his holotype cranidium of *J. missouriensis*. Ross figured what he considered to be the correct pygidium (Ross, 1951, pl. 15, fig. 15), which is clearly a *Peltabellia* species, but one having a relatively flat border and narrow pleural fields compared with *P. glabra*. It is not clear to us that the pygidia figured by Cullison (1944) that Lochman (1966) attributed to *P. willistoni* are the same taxon, as they do appear to have narrower borders than Lochman's types. However, it is clear that they lack a fourth pleural rib, and they are not, therefore, conspecific with *P. glabra*. A pygidium figured by Boyce (1989, pl. 29, figs 3–5) from the Barbace Cove Member of the Boat Harbour Formation, western Newfoundland, under the name *Peltabellia* cf. *P. peltabella* (Ross) appears somewhat similar to that of *P. glabra*, although the latter has a more distinct third pair of pleural furrows. Loch (2007) described a new species as *Strigigenalis implexa* from the Kindblade Formation of Oklahoma and placed Boyce's specimen into its synonymy. The pygidium of Loch's species is similar to that from Newfoundland and that of *Peltabellia glabra*; however, specimens illustrated by Loch (2007, pl. 18, figs 5–8) all show a distinct concavity at the posterior margin, which is not seen on our material, nor, apparently on Boyce's. The Kindblade material also shows a very short pair of posterior pygidial pleural furrows defining a third rib, while the cephalic anterior and genal borders are effaced on the Spitsbergen specimens, but well defined on the Kindblade ones; the marginal raised ridges on the free cheek are apparently unique to *P. glabra*.

Genus *Petigurus* Raymond, 1913

?2007 *Speyeris* Loch, pp. 50–51.

Type species. – Bathyurus nero Billings, 1865, St George Group, western Newfoundland, by monotypy.

Discussion. – Diagnoses of the genus have been provided by Boyce (1989) and Loch (2007). Apart from typically being of large size compared with most bathyurines, *Petigurus* has a long (sag.) glabella with a forwardly protruding frontal lobe that overhangs the narrow anterior border. The glabella may or may not expand in width forwards. Sharply recurved, well-defined palpebral lobes at cephalic mid-length are typical. Stout genal spines have been included in the diagnosis, but it is now known that *P. cullisoni* Loch, 2007, has short ones. Loch (2007) also proposed a new genus *Speyeris* on the basis of disarticulated material from the Kindblade Formation, Oklahoma. The cranidium, including that of the holotype of the type species *S. hami*, looks identical to that of *Petigurus* apart from finer tuberculation (e.g. Loch 2007, pl. 25, figs 8–12), even to the extent of having the typically laterally pointed interocular cheek. However, the pygidium assigned (Loch 2007, pl. 14, figs 9, 10) is different, because it has a wide border that is not steeply inclined beneath the pleural fields, and the border itself is crossed by radiating furrows. The structure of the pygidial border is more reminiscent of *Bolbocephalus kindlei* Boyce, 1989, or even a bathyurelline like *Punka* in which the pleural and interpleural furrows have become particularly deep. It might be possible to relate it to the well-furrowed pygidium of *Petigurus groenlandicus* described below, but one in which there is a curious angular disjunction between the third and fourth pleural segments – in which case one would assume that the presence of the border was a retained primitive feature. It is conceivable that the pygidium does not really belong and indeed is from a different collection to the cranidia. In our view, *Speyeris* is likely to be a subjective synonym of *Petigurus*, based on the type material of the former.

Petigurus nero (Billings, 1865)

Figure 15

1865 *Bathyurus nero* Billings, pp. 260–261, fig. 243.

1913 *Petigurus nero* (Billings); Raymond, pp. 58–59, pl. 7. fig. 8.

1953 *Petigurus nero* (Billings); Whittington, p. 658, pl. 66, figs 18–20, 23–25.

1979 *Petigurus nero* (Billings); Fortey, pp. 84–86, pl. 29, figs 1–12,15.

1986 *Petigurus nero* (Billings); Fortey, p. 19, pl. 2, figs 2, 6–7.

1992 *Petigurus nero* (Billings); Fortey, p. 118, fig. 1a, e.

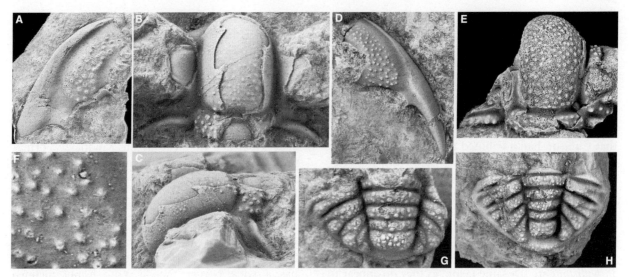

Fig. 15. Petigurus nero (Billings, 1865). Nordporten Member, *P. nero* fauna, 150–200 m. **A**, Free cheek CAMSM X 50188.19 in dorsal view (x1.5). **B–C**, Small cranidium CAMSMX 50188.80 in dorsal and lateral views (x5). **D**, Free cheek CAMSM X 50188.30 in dorsal view (x3). **E**, Enlargement of **D** showing sculpture (x10). **F**, Cranidium CAMSM X 50188.31 in dorsal view (x1.5), ca. 180 m. **G**, Small stage of pygidium CAMSM X 50188.26 in dorsal view (x8). **H**, Large pygidium CAMSM X 50188.32 in dorsal view (x2) ca. 190 m.

2010 *Petigurus nero* (Billings); Boyce and Knight, pl. 2, fig. c.

Material. – Abundant material includes cranidia, CAMSM X 50188.31, X 50188.80; free cheeks, CAMSM X 50188.19, X 50188.30; pygidia, CAMSM X 50188.32, X 50188.26.

Stratigraphic range. – Upper part of Nordporten Member of Kirtonryggen Formation. From 150 to 200 m.

Discussion. – The type species of *Petigurus* from the Catoche Formation, western Newfoundland, was described fully by Fortey (1979), and a repetition of that description is not necessary here. *P. nero* is common in the upper Nordporten Member and is illustrated here for comparison with the material from western Newfoundland. Typical of the species are the lobe-like inflated areas on the outer edge of the pygidial pleurae, which distinguish this species from the older *P. cullisoni* Loch, 2007, and *P. groenlandicus* Poulsen, 1937, and the younger *P. inexpectatus* Fortey & Droser, 1996. The very stout, blade-like genal spines and unusually coarse tuberculate sculpture are also characteristic. We figure here (Fig. 14B, C) a small cranidium of *P. nero* for the first time, at which size it is more similar to that of *Jeffersonia*. The coarse sculpture is still present, but sparser anteriorly, and the profile of the glabella and smaller palpebral lobes serve to distinguish it from *Jeffersonia* spp. of similar size. There are indications

of smooth patches on the flanks of the glabella representing muscle attachment areas, which are lost later. As in the type locality, a small pygidium has a relatively wide axis and the divided pleural lobes are more distinctly drop like than in larger specimens. *P. nero* is a useful taxon stratigraphically because it has proved relatively widespread. It is known from North Greenland (Fortey 1986) and northwest Scotland (Fortey 1992) in addition to its Spitsbergen and Newfoundland occurrences. It is interesting that *P. nero* has not yet been reported from the western United States, possibly because the carbonate platform facies with which it is associated is not developed there. *Petigurus* is one of the largest bathyurines and has the thickest cuticle of any species we know.

Petigurus groenlandicus Poulsen, 1937

Figure 16

1937 *Petigurus groenlandicus* Poulsen, pp. 49–50, pl. 6, figs 1–3, 6–13; ?figs 4, 5.

Material. – Includes cranidia, PMO 208.124, 208.226, 221.253, 223.125; free cheeks, PMO 208.155, 221.256; pygidia, PMO 208.152, 221.257, 223.177; hypostome, PMO 221.260; thoracic segment, PMO 223.250.

Stratigraphic range. – Lower to mid-part of the Nordporten Member, 0–130 m from base, *Petigurus*

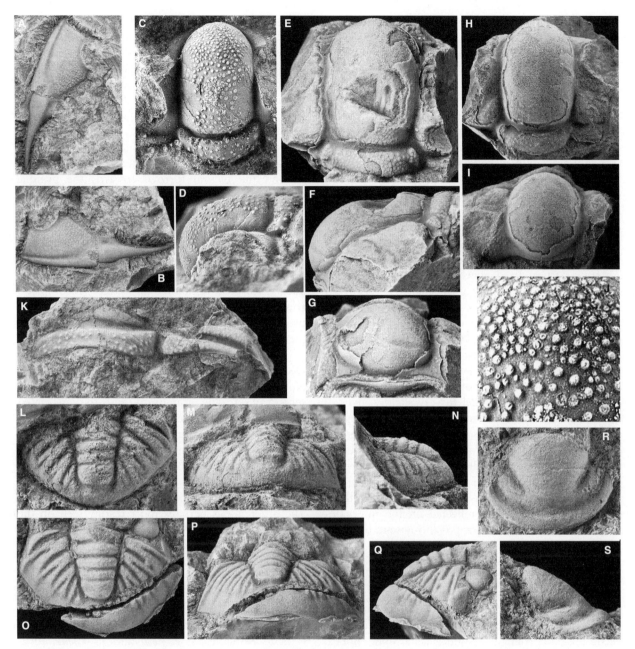

Fig. 16. *Petigurus groenlandicus* Poulsen, 1937. Nordporten Member, *P. groenlandicus* fauna. **A–B**, Free cheek PMO 221.256 in dorsal and lateral views (x1), boundary beds including top 2 m of Bassisletta Member and basal Nordporten Member. **C–D**, Small cranidium PMO 223.125 in dorsal and lateral views (x6), 105 m above base of member. **E–G**, Large cranidium PMO 208.226 in dorsal, lateral and anterior views (x2.5), 130 m. **H–I**, Large cranidium with more subdued sculpture, PMO 208.124 in dorsal and anterior views (x2.5), 105 m. **J**, Enlargement of **C** showing sculpture (cf. fig. 15E). **K**, Thoracic segment PMO 223.250 in dorsal view (x2), boundary beds including top 2 m of Bassisletta Member and basal Nordporten Member. **L–N**, Pygidium with cuticle PMO 223.177 in dorsal, posterior and lateral views (x3), 105 m. **O–Q**, Internal mould of large pygidium 221.257 in dorsal, posterior and lateral views (x1.5), boundary beds including top 2 m of Bassisletta Member and basal Nordporten Member. **R–S**, Hypostome PMO 221.260 in dorsal and lateral views (x4), 105 m.

groenlandicus fauna (Ibexian, early Blackhillsian), probably overlapping with topmost *Chapmanopyge* fauna.

Diagnosis. – *Petigurus* species generally similar to the type species; but glabella often gently forwardly expanding and with comparatively finer tuberculate sculpture on axis and cheeks; large pygidium with well-developed, relatively straight interpleural and pleural furrows.

Description. – Large convex species, with glabella up to about 5 cm long and thick cuticle, so that specimens exfoliate easily. Glabella transversely convex, but with sag. convexity confined to anterior end, where it bulges forwards over anterior border. From in front of occipital ring, glabella expands in width forwards, to a maximum which is 60–67% total sagittal length in dorsal view. Of this, the occipital ring is <20%. Glabellar furrows absent other than a suggestion of furrows opposite palpebral lobes on low flanks of glabella where tuberculation is reduced. Occipital ring as wide transversely as widest part of pre-occipital glabella. Lateral lobes on occipital ring weakly indicated. Furrow defining the ring is deep, transverse, widening medially. Furrow circumscribing glabella deep and uniform. Palpebral lobes of moderate length for a bathyurid, mid-point at about 30% glabellar length, with deeply defined palpebral rims; interocular cheek pointed into rim. Anterior part of fixed cheek downsloping steeply, narrow (tr.) and widening only slightly anteriorly. Posterior part of fixed cheek sloping down less steeply, wide (tr.), with deep border furrow and very convex posterior border. Pre-glabellar field very short, forming a vertical band tucked beneath anterior glabellar lobe. Anterior border weakly defined and narrow. Facial sutures of usual bathyurid form, anterior sections sub-parallel to very slightly divergent in dorsal view, curving adaxially at front of glabella; posterior sections more highly divergent, cutting posterior margin at acute angle. Surface sculpture of tubercles, smaller than those of *P. nero*; occipital ring has room for three to four transverse rows very loosely arranged. There is variation in the extent to which they are reflected on the internal moulds, and they are weaker, possibly absent, on the anterior of the glabella.

Free cheek known from a better specimen than available to Poulsen (1937) carrying similar sculpture to that on glabella, on gently convex genal field. Deeply defined lateral border narrowing towards anterior edge, with round cross section, becoming more triangular towards genal spine. Border furrow deep, curving adaxially inwards at genal spine. The latter is very strong and straight with rounded sub-triangular cross section, with depressed area on inside edge where meeting posterior border furrow. Thoracic segment with band-like axial ring carrying tubercles similar to glabella, prominent articulating half-ring, pleura distally downturned where facet developed, obliquely crossed by exceedingly deep pleural furrow, fading distally; posterior pleural margin with low ridge where abutting segment behind.

Pygidium with rounded triangular outline, length about 80% maximum width near anterior edge. Axis at its widest anteriorly where it is the same width (tr.) as adjacent pleural field, slightly

funnel shaped, tapering most at first ring and close to parallel sided at terminal piece. The four rings are equal sag., though of decreasing convexity posteriorly, the rounded terminal piece twice as long. Axial furrow deep, but furrow around terminal piece shallower, but still discernable. The four pleural segments are all clearly defined, but shorter and more posteriorly sloping backwards; pleural furrows are deeper than interpleural furrows, which are deepest medially, and extend beyond the pleural furrows towards the pygidial margin. Between the fourth pleural segment and the axis, there is a slightly inflated triangular area, somewhat drop shaped. Area behind the axis is also slightly inflated. Border is almost entirely lost, just a weak indication of change in slope at outer ends of pleural furrows. Tuberculate sculpture allows for two (mostly) or three (incomplete) rows of tubercles on axial rings, and they are present on terminal piece and convex bands of pleural areas, but feebly if at all on border area.

Discussion. – Material from the mid-part of the Nordporten Member compares exactly with Poulsen's (1937) type cranidium and the associated pygidium from the Kap Weber Formation, Greenland. Another cranidium figured by Poulsen (1937, pl. 6, fig. 4) has somewhat courser tuberculation and a more forwardly expanding glabella. A difference from *P. nero* is the finer tuberculation on *P. groenlandicus*, although this is a little variable. The difference is perhaps most readily seen on the fewer rows of tubercles on the occipital and pygidial rings in *P. nero*. The genal spine of *P. groenlandicus* is straighter. The pygidium of *P. groenlandicus* displays sub-equal development of the pleural and interpleural furrows and lacks the lobe-like inflated areas towards the edge of the pygidial pleural field which can be used to identify even fragments of *P. nero*. Nonetheless, *P. nero* and *P. groenlandicus* are clearly closely related, and both *P. cullisoni* Loch, 2007, and *P. inexpectatus* Fortey & Droser, 1996, have very different pygidia that suggest a more distant relationship. Cranidia of *Petigurus* sp. 1 Loch, 2007, ranging from the *B. rhochmotis* to *P. cullisoni* zones of the Kindblade Formation, Oklahoma, resemble smaller specimens of *P. groenlandicus*, but the definitive pygidium has not been collected there.

Genus *Psalikilopsis* Ross, 1953

Type species. – *Psalikilopsis cuspicaudata* Ross, 1953, Garden City Formation, western USA.

Remarks. – The type species was hitherto known from relatively small-sized silicified material, and as with the bathyurid *Licnocephala* (p. 21), this makes for some problems in assigning material from other localities to the genus. However, Adrain *et al.* (2011b) have now reviewed the genus from beautifully preserved silicified material from western USA. It is of interest that a species we refer to *Psalikilopsis* occurs in Spitsbergen, although the lack of a pygidium prevents us from naming it formally.

Psalikilopsis n. sp. aff. *P. cuspicaudata* Ross, 1953

Figure 33A–K

Holotype. – Cranidium, PMO 223.185 (Fig. 33I–K); Nordporten Member, 105 m.

Material. – Cranidia, PMO 223.164, 223.173, 223.184, 223.245; free cheek, PMO 221.262, 223.227.

Stratigraphic range. – Nordporten Member, 105–130 m from base, *Petigurus groenlandicus* fauna.

Description. – Although we have only cranidia and cheeks of this species, the rarity of the genus elsewhere and its occurrence in Spitsbergen justify a description in this particular case. It is a small species, cranidium not exceeding 5 mm; cuticle not particularly thick, so that sculpture is reflected on internal moulds. Cranidium highly convex, both sag. and trans., with glabella bulging forwards to hardly or slightly overhang pre-glabellar field. In dorsal view, glabella with ovoid outline in front of occipital ring, and length (sag.) is about 1.5 times maximum width at palpebral lobes. In anterior view, the outline of the frontal lobe is almost circular. Axial furrows deep and rather uniform along their length, evidently wider on internal moulds. Occipital furrow slightly narrowing medially, and one specimen shows indication of slightly swollen areas *within* the furrow to either side of this. Occipital ring narrow (sag., exsag.) 0.15 total glabellar length (dorsal view). Posteromedial occipital spine is short and pointed and directed upwards and slightly backwards. In front of it is an occipital tubercle; it is apparently little different from other exoskeletal tubercles, but it is consistently present in what is otherwise a smooth area. Palpebral lobe very close to glabella and one-third (exsag.) the length of it (dorsal aspect), having comparatively gently curved outline describing a short arc of a circle, rim well defined by palpebral furrow, flattened compared with somewhat inflated

genal area inside it. Posterior fixed cheeks are acutely triangular, defined by highly divergent posterior limbs of the facial sutures, and are at least two-thirds (tr.) as long as the occipital ring. The deep posterior border furrow divides the convex anterior portion of the fixed cheek from a border of rather uniform width along its length. The anterior section of the facial suture is bowed outwards in front of the palpebral lobe, hence curving gently adaxially to cut the anterior border at approximately a right angle. The anterior part of the fixed cheeks and moderately wide pre-glabellar field slope downwards steeply until abruptly terminated at anterior border furrow. Convex, ledge-like anterior border bowed upwards steeply about mid-line, border furrow forming an inverted V shape. Mid-part of pre-glabellar field may show obscure median (sag.) depression. Sculpture consisting of distinctively scattered tubercles extending over mid-part of glabella that are smaller and sparser posteriorly and anteriorly. Tubercles absent on occipital ring at least medially and on pre-glabellar area and border. The same scattered sculpture enables us to associate a free cheek, which would have sloped down very steeply along the lateral edge of the cephalon, forming rather more than a quarter circle in outline. One specimen has broken along the doublure, which is narrowly reflexed upwards to create a marginal cephalic tube in continuity with the cranidial border. Cheek continues into narrow and pointed genal spine, which must have been nearly as long (exsag.) as the genal field.

Discussion. – The illustration of this species in Fig. 33 is out of sequence because we originally assigned it to family Dimeropygidae. At a late stage in the production of this work, we were made aware of a paper by Adrain *et al.* (2011b) which transferred *Psalikilopsis* to the family Bathyuridae. It may be remarked that we only discovered *P.* n. sp. aff. *P. cuspicaudata* by breaking up fossiliferous limestones in the laboratory from the productive trilobite 'hash' beds of the 105-m horizon in the Nordporten Member. It is scarcely possible to recognise it in the field. It seems possible that other, small-sized genera may be present in the Spitsbergen Early Ordovician, but more bulk sampling (or the discovery of silicification) will be necessary to discover them.

Like other species attributed to *Psalikilopsis*, the species does not attain a large size. Among the species described in detail by Adrain *et al.* (2011b), the Tulean type species, *P. cuspicaudata*, has a very markedly arched cranidial border like the species from the Kirtonryggen Formation. Stratigraphically earlier species attributed to the same genus have a less vaulted anterior outline. However, the species from the Nordporten Member has in addition a sharp occipital spine, dorsally directed, which does not appear to be present on any of the species from the Great Basin. It therefore seems that a new *Psalikilopsis* species is present in Spitsbergen. Pygidia are particularly important in the taxonomy of the species according to Adrain *et al.* (2011b), but we have not been able to find one by breaking up samples. The best recourse seems to be to the use of open nomenclature for the moment, and we opt to compare the present species with *P. cuspicaudata.*

Bathyurine n. gen., n. sp. A

Figure 17

Material. – Cranidia, PMO 208.220, CAMSM X 50188.34.

Stratigraphic range. – Collected *in situ* 130 m from base of Nordporten Member, upper part of range of *Petigurus groenlandicus* fauna.

Description. – Cranidium with very steep anterior slope, length in dorsal view 70% or more maximum transverse width at posterior margin. Glabella parallel sided and unusually long and sub-rectangular (width in dorsal view 60–70% length) not greatly vaulted transversely with faint median crest. In ante-

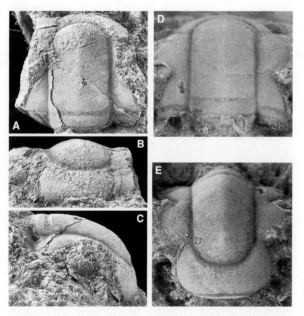

Fig. 17. Bathyurine n. gen., n. sp. A. Nordporten Member, *P. groenlandicus* fauna. **A–C,** Cranidium CAMSM X 50188.34 in dorsal, anterior and lateral views (x4), ca. 130 m. **D–E,** Cranidium PMO 208.220 in dorsal and anterior views (x5), 130 m.

rior view, the front of the glabella is rather bluntly rounded, and axial and pre-glabellar furrows of even definition not deep. Glabellar furrows lacking, but occipital furrow of similar depth to axial furrows defining narrow (sag.) occipital ring one-sixth or less total glabellar length. Small median occipital tubercle on internal mould. Palpebral lobes positioned forwards for a bathyurid and well removed from glabella, and length (exsag.) slightly <40% sagittal glabellar length, with arcuate outline and weakly defined narrow rims. Posterior sections of facial sutures are unusually weakly divergent behind the eyes, making an angle of about 30° to the sagittal line, and cutting the border at a high acute angle, while the anterior sections are similarly weakly divergent and curved uniformly adaxially as they pass downwards towards the cranidial margin, which they cut opposite the axial furrows. Posterior fixed cheeks are shortly triangular (tr.), the distinct border furrow arched gently forwards, outlining a rather flat posterior border about half as wide (exsag.) as the occipital ring (sag.). The anterior cranidial border, by contrast, is very narrow and convex, defined by a narrow and deep border furrow. Above it, the wide pre-glabellar field slopes down close to vertically, such that it is much foreshortened in dorsal view. Although much material is exfoliated, there is no indication of tuberculation on the internal mould, and such pieces of exoskeleton as are retained are entirely smooth, so it is likely that surface sculpture was lacking.

Discussion. – This trilobite deviates from general bathyurid morphology in having forwardly placed eyes and narrowly triangular cheeks behind the eyes, defined by weakly divergent sutures. There is possibly some comparison with *Harlandaspis* in its long glabella, but the rectangular glabella shape and the long, steeply sloping pre-glabellar field of the present species are distinctively different. We did consider the possibility of it belonging to another family altogether, but there is probably no other plausible option. The species is likely to belong to a new genus, but we have not been able confidently to associate a free cheek or pygidium. The closest match among described trilobites assigned to Bathyuridae is superficially with *Gelasinocephalus pustulosus* Loch, 2007, from the Kindblade Formation, Oklahoma, which shares with our species the steep anterior cranidial profile, vertical pre-glabellar field and narrow cranidial border and also has triangular post-ocular cheeks. With its coarsely tuberculate sculpture and deeply defined tapering glabella, it is equally also clearly different from the Spitsbergen form. *G. pustulosus* is confusingly also very different from the type species of *Gelasinocephalus*, *G. whittingtoni* Loch, 2007, which has a smooth, forwardly protruding, expanded glabella, narrower (exsag.) posterior fixed cheeks and palpebral lobes in the usual bathyurid position. It is has been associated with a free cheek (Loch 2007, pl. 9, figs 12,13) closely similar to those assigned to *Bolbocephalus* elsewhere (e.g. Fig. 6T), and the pygidium is like that of *Rananasus*, a relative of *Bolbocephalus* (p. 21), which has been regarded as its junior synonym (Adrain & Westrop 2005). A free cheek associated with *G. pustulosus* (Loch 2007, pl. 10, figs 7, 8) by contrast carries very strong raised ridges on the border and has tubercles of two sizes, both features suggestive of dimeropygid affinities (Adrain *et al.* 2001). In fact, its cranidium is also like that of several *Ischyrotoma* species with relatively effaced anterior cranidial borders. No pygidium was recognised for *G. pustulosus* that would have proved its family affinities. Our view is that the type species of *Gelasinocephalus* is related to *Bolbocephalus*, from which a longer (sag.) pre-glabellar field possibly distinguishes it, and that *G. pustulosus* is not related to it. The species under discussion from Spitsbergen is related to neither. Its long, rectangular glabella and the form of its palpebral lobes (and indeed the relatively large overall size) preclude dimeropygid relationships, the triangular posterior fixed cheeks being a convergent similarity to members of that family. Its novel features might require generic recognition, but it would be unwise to add to an already confused situation by proposing one here, before the appropriate free cheek and pygidium have been recognised. Hence, the use of open nomenclature herein.

Sub-family Bathyurellinae Hupé, 1953

Diagnosis. – Bathyurids with relatively large, frequently flattened pygidia with wide borders and with equally wide doublure with inner edge usually reflexed close to dorsal surface. Glabella parallel sided, sometimes acuminate anteriorly. Free cheeks falcate, with genal spines broadly an extension of genal fields. Dorsal surface sculpture either lacking or typically with terrace ridges, not tuberculate.

Remarks. – Although sub-family classification of bathyurids is still debatable, it is very likely that the species with large pygidia centred on *Bathyurellus* do comprise a clade. Hence, Bathyurellinae certainly includes *Bathyurellus*, *Punka*, *Uromystrum*, *Licnocephala*, *Grinnellaspis* and *Chapmanopyge* n. nom. (*pro Chapmania* Loch, 2007). These constituent genera

are discussed further below. For the moment, we also include *Harlandaspis* n. gen. and *Benthamaspis* in the same sub-family, although they lack the typical flattened pygidium and probably together comprise a natural group. This is because an early species assigned here to *Harlandaspis*, *H. inflatus* (Loch, 2007), shows features that link the cephalon to *Bathyurellus*, to which genus it was assigned by Loch (2007). *Benthamaspis* has moved even further away from typical bathyurelline morphology.

Pending clarification of type species of *Grinnellaspis* and *Licnocephala* (see below), a practical classification based on pygidia of the typical bathyurellines might be:

Licnocephala: flat border on pygidium, not deeply marked at inner edge, pleural furrows at most weakly cross on to it. Overall convexity very low, often matched by wide cephalic border.

Punka: pygidial border not deeply demarcated from pleural fields, sigmoidal furrows cross on to it and are often quite wide.

Grinnellaspis: pygidial border not deeply demarcated from pleural fields, pleural furrows narrow, straight, long and radially arranged.

Chapmania Loch 2007 (type species *C. oklahomensis* Loch, 2007; now *Chapmanopyge* herein, see p. 63) includes stratigraphically early members of this group, some of which have sub-rectangular glabellas. This seems to be the plesiomorphic condition for bathyurellines, if not all bathyurids, as seen on *Peltabellia glabra* (above). The pygidia of species assigned to *Chapmania* show sharply defined pygidial borders with at most weak pleural furrows crossing on to them. Loch (2007) placed *Licnocephala sminue* Fortey & Peel, 1990, from Greenland into *Chapmania*. There are often comparably sharp border furrows on the cephalon, which provides a discriminating feature from *Licnocephala* as understood herein.

Genus *Bathyurellus* Billings, 1865

Type species. – *Bathyurellus abruptus* Billings, 1865, Catoche Formation, western Newfoundland.

Discussion. – *Bathyurellus* was formerly applied in a wide sense to embrace bathyurids with asaphid-like pygidia, but following revision of the type species, it is used in a more restricted way for bathyurellines having pygidia with narrowly convex adaxial pleural fields and broadly sloping borders giving the whole a resemblance to a duck's bill. The V-shaped inner outline of the pygidial doublure also appears to be characteristic (see remarks under *Punka*, below).

Bathyurellus abruptus Billings, 1865

Figure 18

1865   *Bathyurellus abruptus* Billings, pp. 263–264, figs 247, 250.

1905   *Bathyurellus abruptus* Billings; Raymond, p. 337.

1953   *Bathyurellus abruptus* Billings; Whittington, p. 661, pl. 69, figs 26–27.

1979   *Bathyurellus abruptus* Billing; Fortey, pp. 92–94, pl. 32.

1986   *Bathyurellus abruptus* Billings; Fortey, pp. 19–20, pl. 2, figs 2, 6, 7.

2010   *Bathyurellus abruptus* Billings; Boyce and Knight, pl. 2, fig. F.

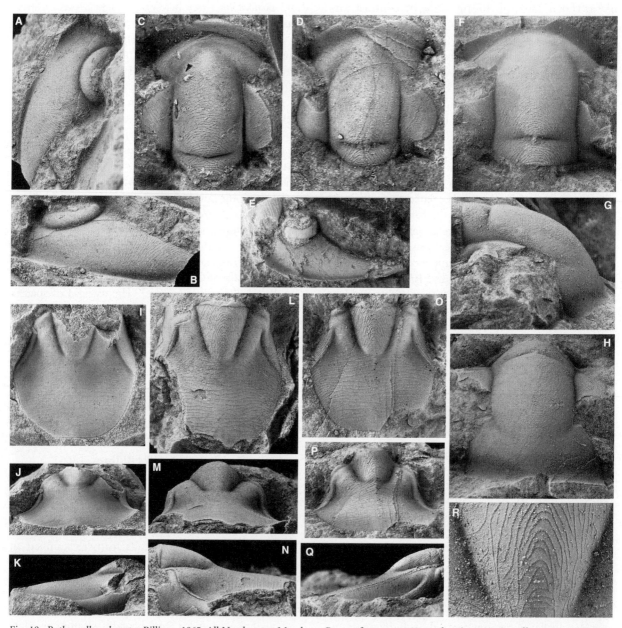

Fig. 18. Bathyurellus abruptus Billings, 1865. All Nordporten Members, *P. nero* fauna ca. 160 m, showing anterior effacement. **A–B**, Free cheek CAMSM X 50188.47b in dorsal and lateral views (x5). **C**, Cranidium CAMSM X 50188.48 in dorsal view (x7). **D**, Cranidium CAMSM X 50188.23 in dorsal view (x8). **E**, Free cheek PMO 208.199 in lateral view (x3), 190 m. **F–H**, Cranidium CAMSM X 50188.53 in dorsal, lateral and anterior views (x7) showing anterior effacement. **I–K**, Pygidium CAMSM X 50188.22 in dorsal, posterior and lateral views (x3). **L–N**, Pygidium CAMSM X 50188.27 in dorsal, posterior and lateral views (x4). **O–Q**, pygidium CAMSM X 50188.29 in dorsal, posterior and lateral views (x7). **R**, Enlargement of **L** showing terrace sculpture (x15).

Material. – Cranidia, CAMSM X 50188.23, X 50188.48, X 50188.53; free cheeks, PMO 208.199; CAMSM X 50188.47b; pygidia, CAMSM X 50188.22, X 50188.27, X 50188.29.

Stratigraphic range. – Nordporten Member, upper part 150 to 200 m, with *Petigurus nero* fauna.

Discussion. – Two species of *Bathyurellus sensu stricto* were described by Fortey (1979) from western Newfoundland: *B. abruptus* Billings, 1865, and *B. platypus* Fortey, 1979, the second being the stratigraphically younger. *B. abruptus* was distinguished from the latter in having a surface sculpture on the glabella of terrace ridges rather than pits and a more concave pygidial margin. Material from the Nordporten Member shows no differences from *Bathyurellus abruptus* from the type locality in the Catoche Formation. Since a full morphological description was given by Fortey (1979), it does not need to be repeated here; however, diagnostic sclerites of the species are illustrated here as a matter of record. The same species is present in the Wandel Valley Formation from eastern North Greenland (Fortey 1986), so this is evidently another of those Blackhillsian species of utility for stratigraphic correlation along the margin of Ordovician eastern Laurentia. In Spitsbergen, it is underlain stratigraphically by a new species, *B. diclementsae*, with a pygidium of different proportions, and a better-defined glabella.

Bathyurellus diclementsae n. sp.

Figure 19

1937 *Bathyurellus teicherti* Poulsen, pl. 7 (*pars*), fig. 5, *non* figs 2–4.

Holotype. – Pygidium, PMO 223.156 (Fig. 19U–W); Nordporten Member, 105 m.

Material. – Among numerous other specimens, cranidia, PMO 223.150, 223.155, 223.186, CAMSM X 50188.39; free cheeks, PMO 223.159; pygidia, PMO 208.219, 208.147, 223.145, PMO NF 1916, NF 1908, CAMSM X 50188.55.

Stratigraphic range. – Lower part of Nordporten Member base to 130 m, with *Petigurus groenlandicus* fauna.

Diagnosis. – *Bathyurellus* species with wide, paddle-shaped pygidium, about 80% as long as wide;

glabella relatively well defined and slightly pointed anteriorly; cranidial anterior border equal in width (sag.) to pre-glabellar field behind it in dorsal view.

Etymology. – For Mrs Di Clements who helped so much with the completion of this work.

Description. – Type series is from the 105-m horizon, and the sclerites are confidently associated. Species of medium size for a bathyurid, with pygidia about 1 cm long, and thin cuticle, easily flaking off. Cranidium moderately convex (sag., tr.) with posterior part of occipital ring and glabella close to horizontal on the mid-line, and with the anterior one-third of glabella sloping downwards, with pre-glabellar field continuing a similar slope to the out-turned, but still gently sloping border. Horizontal palpebral lobes are well below the high point of the glabella. The width/length ratio of the glabella varies between 0.6 and 0.8 (transverse width measured at mid-glabellar length) and is widest at the occipital ring, sub-parallel sided or very slightly tapering forwards in front of that to a point opposite the anterior ends of the palpebral lobes, where its rounded corners converge on an obtuse median point. The axial furrows are well defined but not deep, as are the pre-glabellar furrows. Small cranidia appear more rounded anteriorly in dorsal view. The overall shape is much like that of a gothic church window. There is no indication of glabellar furrows. Occipital furrow deep medially, but shallowing laterally. Occipital ring one-fifth or somewhat less total glabellar length (sag.), slightly wider medially where a suggestion of a median occipital tubercle on some specimens. Palpebral lobes large, one-third or slightly more (exsag.) total cranidial length (sag.) and curved more strongly at posterior ends and twice as long as wide (tr.); palpebral rims hardly discernable. Posterior branches of facial sutures slope downwards behind the palpebral lobes making an angle of 70–80° to sag. line and running more or less straight to the posterior margin, thereby defined a narrowly triangular, acutely pointed post-ocular cheek. Straight posterior border furrow, and hardly convex posterior border widens slightly abaxially. Anterior branch of suture diverges at a lesser angle (45–65° to sag. line in dorsal aspect) and slopes quite steeply downwards until turning adaxially, to make rounded anterolateral corners to the pre-ocular cheeks. Genal field here is slightly convex, narrowest medially, where (seen in dorsal view) the pre-glabellar field has about the same width (sag.) as border beyond; in anterior view, the pre-glabellar field is higher than the border, which is

very gently arched upwards medially. Small cranidia have longer pre-glabellar fields and are generally more convex (sag.). Border itself is really no
more than a change in slope, concave outwards,

downward sloping but nearly horizontal at the
anterior-most edge, which is narrowly convex.

Free cheeks generally blade like, length along
adaxial margin (exsag.) about half max. total length,

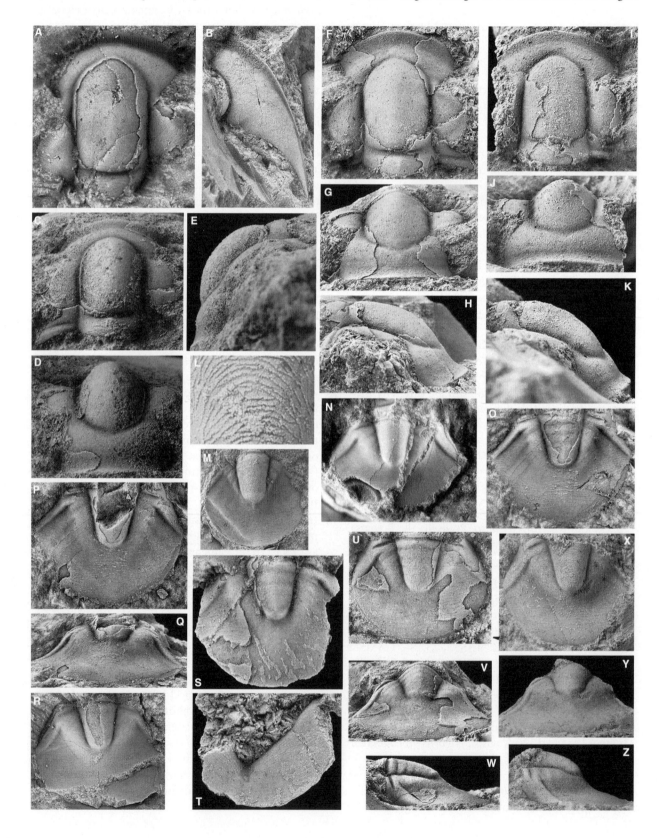

with sharply acute genal angle in dorsal view. Posterolateral parts of cheek are turned close to vertical where the genal corners sweep backwards alongside anterior part of thorax. Lateral border widest where it abuts the cranidium, becoming progressively narrower in a regular fashion all the way to the genal spine, at the lateral edge eventually comprising only the narrowest of rims. Near the posterior margin where the posterior branch of the facial suture cuts across, there is a hint of another furrow that may be the true lateral border (see p. 22). The eye is band like, not very high, and without clearly distinguished socle.

Pygidium has typical *Bathyurellus* shape, that is, with convex, narrow (tr.) adaxial 'shoulders' and concave, sloping flanks flattening out into an indistinctly defined sloping border. However, the pygidium of *B. diclementsae* is broad and paddle shaped, rather than duckbill shaped, and in that regard has more general resemblance to those of other bathyurellines. Length (sag.) lies between 75% and 85% maximum transverse width at posterior edge of facets. Axis is 40% total pygidial length and one-third max width at its anterior edge. Axis tapers regularly to semicircular terminal piece and is well defined throughout, standing clear of rest of pygidium. Four axial rings of sub-equal length (sag.) can be discerned, but only the first ring furrow clearly crosses the mid-part of the axis on internal surface, the rest being faint on the flanks of the axis. Half-ring narrow (sag.). The anterior pleural segment of the pygidium is also clearly defined, with a short, clear, deep, transverse pleural furrow extending across the proximal part of the pygidial field and then stopping. The steep facet cuts backwards at this point, and the posterior edge of the facet is marked by a clear ridge, backed by a furrow sloping outwards and backwards, marking the anterior edge of the border. Pleural field is sub-horizontal in a triangular area adjacent to the axis, laterally and posteriorly sloping downwards steeply before flattening out on the border. A second segment is weakly expressed on the pleural field, just visible on the border. Doublure (Fig. 19T) underlies the sloping border and is horizontal, carrying terrace ridges, with a typical v-shaped internal margin, where it turns dorsally.

Despite its long stratigraphic range, we can detect no differences between the early and late representatives of the species.

Surface sculpture is feeble. Pygidial pleural fields with weak terrace lines, stronger on the axis and glabella, and posterior margin of free cheek; otherwise apparently smooth.

Discussion. – Poulsen (1937) described *Bathyurellus teicherti* from the Kap Weber Formation of northeast Greenland. There is no question that the pygidium attributed to this species by Poulsen is identical to the pygidium of *B. diclementsae* (Fig. 19N herein). However, Poulsen associated the wrong cephalic parts with the pygidium, and the holotype of the species *teicherti* is a cranidium (Poulsen 1937, pl. 37, fig. 2). This cranidium is like that of *Licnocephala* or possibly *Chapmanopyge*, as is the free cheek. Hence, a new name is needed for the true *Bathyurellus* species from Spitsbergen and Greenland described above.

Bathyurellus diclementsae differs from the type species, which it underlies stratigraphically, in its relatively wide pygidium, which has the exaggerated facet typical of the genus but has not fully acquired the distinctive duckbill shape of *B. abruptus*. The glabella is more strongly defined anteriorly, and the surface sculpture appears weaker. *Bathyurellus platypus* Fortey, 1979, from western Newfoundland and New York State (see also Brett & Westrop 1996) is younger again and has similar differences from *B. diclementsae*, with a different cephalic sculpture. Hintze (1953, pl. 9, fig. 9) illustrated but did not describe a pygidium from 'Zone G' in Utah under open nomenclature, which is generally like that of *B. diclementsae*.

One pygidium from the same horizon as *B. diclementsae* (Fig. 19M) shares a similar shape with that species, but has narrower, better-defined triangular areas adjacent to the axis, and clearer, very short pleural furrows. The axis has a short post-axial ridge, and the pygidial border is more distinctly flattened. Although there is a modest amount of variation in the pygidia of *B. diclementsae*, this specimen falls outside it, and it seems likely that there is another much rarer species of *Bathyurellus* in the

---

Fig. 19. Bathyurellus diclementsae n. sp. Nordporten Member, *P. groenlandicus* fauna. **A**, Cranidium CAMSM X 50188.39 in dorsal view (x4), unspecified horizon. **B**, Free cheek PMO 223.159 in dorsal view (x3), 105 m. **C–E**, Small cranidium PMO 223.186 in dorsal, anterior and lateral views (x10), 105 m, showing greater convexity. **F–H**, Cranidium PMO 223.150 in dorsal, anterior and lateral views (x5), 105 m. **I–K**, Cranidium PMO 223.155 in dorsal, anterior and lateral views (x7), 105 m. **N**, Pygidium attributed to *Bathyurellus teicherti* figured by Poulsen (1937, pl. 7, fig. 5) (x5.5) and here assigned to the new species. **O**, Pygidium PMO NF 1916 in dorsal view (x3). **P–Q**, Pygidium PMO NF 1908 in dorsal and posterior views (x3), exact horizon unknown. **R**, Pygidium PMO 208.219 in dorsal view (x3), 70 m. **S**, Pygidium PMO 223.145 in dorsal view (x4), 105 m. **T**, Doublure, underside of specimen in S (x4). **U–W**, Holotype, pygidium PMO 223.156 in dorsal, posterior and lateral views, 105 m. **X–Z**, Stratigraphically early pygidium PMO 208.147 in dorsal, posterior and lateral views (x4), basal Nordporten Member and **L** enlargement of axis showing sculpture. **M**, *Bathyurellus* aff. *diclementsae*. CAMSM X 50188.55 in dorsal view (x3) ca. 100 m. Note narrow, defined pleural fields.

same strata. It cannot be named on the basis of this pygidium alone, and it is referred to provisionally herein as *Bathyurellus* aff. *diclementsae*. A similar pygidium from the Kap Weber Formation, East Greenland, was figured by McCobb & Owens 2008 (pl. 1, fig. 5). Its clarification must wait on further collecting.

Genus *Benthamaspis* Poulsen, 1946

Type species. – *Benthamaspis problematica* Poulsen, 1946, Cape Steven, Ellesmere Island.

Discussion. – *Benthamaspis* was placed in the Family Bathyuridae by Ludvigsen *et al.* (1989) and Boyce (1989), an assignment also followed by Loch (2007). Description of *Harlandaspis* n. gen. (below) adds support to this view, since, unlike *Benthamaspis*, this genus has an elongate glabella more similar to that of other bathyurellines and retains a relatively wide (tr.) free cheek. Like *Benthamaspis,* it has both lost genal spines and acquired a pygidium completely lacking a defined border. Together, *Harlandaspis* and *Benthamaspis* comprise a sub-group within the Bathyuridae, the final status of which will depend on a full phylogenetic analysis of the family. The narrow (tr.) free cheek of *Benthamaspis* is so unlike that of any other bathyurid that its origin requires explanation. It is derived from reduction in the proximal part of the genal field. The early species *Benthamaspis rhochmotis* Loch, 2007, demonstrates the state intermediate with a 'normal' bathyurid since it retains a genal border furrow, very much like that of *Harlandaspis*. In advanced *Benthamaspis* spp., the proximal part of the fixed cheek is reduced to a tiny stub visible under the eye lobe as a short ridge (Fortey 1979, pl. 35, fig. 4). This reduction permits a general narrowing of the exoskeleton to produce a more compact morphology, accompanied by a shortening of the glabella to the sub-quadrate shape typical of the genus.

Benthamaspis includes species ranging through the Lower to Middle Ordovician, from Tulean (Loch 2007) to basal Whiterockian (Fortey & Droser 1999) age. The type species is Blackhillsian, as is the first-described species, *B. gibberula* (Billings, 1865). Adrain *et al.* (2009) have illustrated a series of as yet undescribed silicified species from the Great Basin, western United States, and it would be best to await full description of these species from the type Ibexian before comparing them with our material. As presently conceived, there is a good deal of variation within named *Benthamaspis* species in certain features. The length of the defined occipital furrow can be short to long, for example, and the taper of the pygidial axis varies from almost sub-parallel sided to conical (Fortey 1979; Boyce 1989), with a variably developed post-axial ridge, while the Bertillon sculpture can be more or less dense. More critical definition of species may be possible when large samples of perfectly preserved specimens are available. In general, there is a temporal trend from the stratigraphically early *B. rhochmotis* Loch, 2007, to late *B. gibberula* entailing progressive effacement of dorsal furrows – pre-glabellar and occipital furrows particularly – and loss of pygidial pleural furrows. Although the material from Spitsbergen is well preserved, it is limited in quantity, and we must take a *sensu lato* view of species herein. Given the other unquestionable similarities between the upper faunas of the Kirtonryggen Formation and those of the Catoche Formation, western Newfoundland, it is plausible to discover extended ranges of *Benthamaspis* species between these two localities. As one of the few taxa in common between the Ordovician of the western USA and eastern Laurentia, *Benthamaspis* spp. may prove useful for fine-scale correlation in future. As so often with these Early Ordovician Laurentian faunas, the type species is not well known, and the same caution applies as with other bathyurids. It is, however, certainly very similar to the next species discussed below and may prove synonymous.

Benthamaspis gibberula (Billings, 1865)

Figure 20A–H

1865 *Dolichometopus*? *gibberulus* Billings, p. 269, fig. 254.

?1946 *Benthamaspis problematica* Poulsen, 1946, pp. 325–326, pl. 22, figs 14–16.

1969 '*Bathyurus*' *gibberulus* (Billings); Whittington and Kindle, p. 658.

1979 *Benthamaspis gibberula* (Billings, 1865); Fortey, pp. 100–102, pl. 43, figs 1–15.

1989 *Benthamaspis gibberula* (Billings); Boyce, p. 11.

?1996 *Benthamaspis striata* (Whitfield, 1897); Brett and Westrop, pp. 425–426, fig. 19.10, 19.11.

Material. – Cranidia, PMO 208.197, CAMSM X 50188.6; free cheeks, CAMSM X 50188.15; pygidia, PMO 208.198, CAMSM X 50188.81.

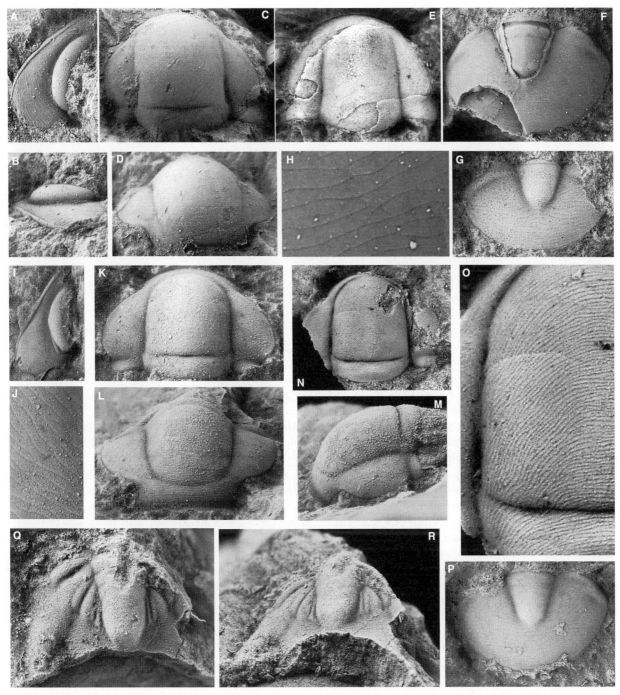

Fig. 20. **A–H**, *Benthamaspis gibberula* (Billings, 1865). Nordporten Member, *P. nero* to *L. platypyga* fauna. **A–B**, Free cheek CAMSM X 50188.15 in dorsal and lateral views (x12), *L. platypyga* fauna, top 3 m of Nordporten Member. **C–D**, Cranidium PMO 208.197 in dorsal and lateral views (x7), upper *P. nero* fauna, 190 m. **E**, Cranidium CAMSM X 50188.6 in dorsal view (x7), *L. platypyga* fauna, top 3 m of Nordporten Member. **F**, Pygidium PMO 208.198 in dorsal view (x6), upper *P. nero* fauna, 190 m. **G**, Smaller pygidium CAMSM X 50188.81a in dorsal view (x13), *L. platypyga* fauna, top 3 m of Nordporten Member. **H**, Enlargement of G (x20) showing the sculpture. **I–P**, *Benthamaspis conica* Fortey, 1979. Nordporten Member, *P. nero* fauna and *L. platypyga* fauna. **I**, Free cheek PMO 221.233 in dorsal view (x5), 190 m, showing sub-ocular ridge. **J**, Enlargement of A showing the sculpture (x15). **K–M**, Cranidium CAMSM X 50188.17 in dorsal, anterior and lateral views (x12), probably from top 3 m, *L. platypyga* fauna. **N**, Cranidium CAMSM X 50188.18 in dorsal view (x6) with (**O**) enlargement (x15) showing the sculpture and long occipital furrow, *P. nero* fauna, unspecified horizon. **P**, Pygidium PMO 208.098 in dorsal view (x9), 150 m. **Q–R**, *Ceratopeltis* cf. *C. batchensis* Adrain & Westrop, 2005, Topmost Bassisletta Member, top of *Chapmanopyge* fauna. Pygidium PMO 223.228 in dorsal and posterior views (x11).

Stratigraphic range. – Nordporten Member, top 30 m, upper *Petigurus nero* fauna and *Lachnostoma platypyga* fauna.

Discussion. – A full description of this species based upon topotype material from the Catoche Formation western Newfoundland was given by Fortey (1979), who recognised that Billings' (1865) species *Dolichometopus? gibberulus* should be assigned to *Benthamaspis*. Additional descriptions of *Benthamaspis* species by Boyce (1989) and Brett & Westrop (1996) agree in many general features, and a full description does not require repetition here. Well-preserved specimens from the top of the Kirtonryggen Formation conform to the diagnosis of *B. gibberula* given in Fortey (1979). On the cranidium, in particular, effacement of the pre-glabellar furrow was regarded as an important character to distinguish this species from *B. conica* Fortey, 1979. This feature is clear on a specimen from the higher part of the Nordporten Member in Spitsbergen (Fig. 20E), which also has a sub-rectangular glabella and weakly visible palpebral rims, features which can be matched by specimens used to illustrate *B. gibberula* from Newfoundland (e.g. Fortey 1979, pl. 34, figs 2, 4, 7). It also seems typical that the pre-glabellar field slopes downwards so steeply as to be invisible in dorsal view. However, there is variation in the depth and extent of the occipital furrow, which is longer (tr.) in the Spitsbergen specimen closest in other ways to figured material from the Catoche Formation. But other Catoche cranidia attributed to *B. gibberula* have longer occipital furrows (e.g. Fortey 1979, pl. 34, fig. 1), so at the moment, such variation is regarded as within the species. This view might be reinforced by a cranidium from the Kirtonryggen Formation (Fig. 20F) on which the occipital furrow is almost completely effaced; the palpebral lobes are also less curved on this specimen. A free cheek from the top of the Kirtonryggen Formation is of interest as it lacks the posterior sub-ocular ridge that is the last remnant of the genal field (see above). The final effacement of this ridge might be regarded as an even more advanced character than its reduction to a stub, and it does differ from the cheek illustrated by Fortey (1979, pl. 34, fig. 6) for *B. gibberula* in this detail and may be a more plausible association given the greater effacement of *B. gibberula* generally. The type and only specimen of *B. striata* (Whitfield, 1897) from the Fort Cassin Formation, Vermont, is a cranidium reillustrated by Brett & Westrop (1996), and it, too, shows the effaced pre-glabellar furrow typical of *B. gibberula*; the occipital furrow is of the long type. It might prove to be a junior synonym of Billings' species, although a ques-

tion remains without its pygidium and free cheek. Given the other species similarities between the Catoche Formation and the Fort Cassin Formation, this outcome is likely. Differences from other species of *Benthamaspis* are considered further under *B. conica* (below).

Benthamaspis conica Fortey, 1979

Figure 20I–P

1966 *Oculomagnus obreptus* n. gen., n. sp. Lochman, pp. 541–542 (*pars*), pl. 62, figs 1, 2, 4.

1979 *Benthamaspis conica* sp. nov. Fortey, pp. 102–104, pl. 35, figs 1–10.

1986 *Benthamaspis conica* Fortey; Fortey, pp. 20–21, pl. 1, figs 1, 2.

1989 *Benthamaspis hintzei* sp. nov. Boyce, pp. 56–57, pl. 32, figs 1–8.

2011 *Benthamaspis conica* Fortey; McCobb *et al.*, fig. 2E.

Material. – Cranidia, CAMSM X 50188.17–18; free cheek, PMO 221.233; pygidium, PMO 208.098.

Stratigraphic range. – Upper 70 m of Nordporten Member, *P. nero* fauna to *L. platypyga* fauna.

Discussion. – This species has been described by Fortey (1979) and discussed further by Fortey (1986), Boyce (1989) and Brett & Westrop (1996); most of its features are shared with the preceding species, so repetition is avoided here. A cranidium fitting within the range of variation of the type series shows clearly the well-defined anterior margin of the gently tapering glabella that distinguishes the species from *B. gibberula*. The long, narrow occipital furrow that almost reaches the axial furrow before shallowing is also typical, although as noted above this feature is apparently variable in *B. gibberula*. In dorsal view, the pre-glabellar field is visible because the slope of the pre-glabellar field is not as steep as typical for *B. gibberula*, though this, too, appears to be variable. Figure 20K compares exactly with the holotype (Fortey 1979 pl. 35, fig. 1) in this regard. Boyce (1989) erected a species (*B. hintzei* Boyce, 1989) from the same part of the Catoche Formation, western Newfoundland, as that yielding *B. conica*. The type material of *B. hintzei* is imperfectly preserved, but cranidia show the long occipital furrow and an anteriorly defined glabella, in front of which the pre-glabellar field is visible (Boyce 1989, pl. 32, fig. 3), both defining features of *B. conica*. Boyce's type is a pygidium, which is distinguished (Boyce 1989, p. 57) from that of *B. conica*, and is stated to have 'axial ring furrows strongly anteriorly convex'. Like almost all bathyurids, *Benthamaspis* spp. have four pygidial segments reflected as transverse furrows on the axis, plus a terminal piece. In *B. conica,* they are effaced dorsally, but still visible as transverse furrows on internal moulds (cf. Boyce 1989, pl. 32, fig. 4). The 'anteriorly convex' ring furrows are actually a variable expression in the surface sculpture of equivalents to the prominent axial half-ring that is present at the anterior margin of the pygidium in all species of the genus. It is possible to make out this feature on the front of the second axial ring of the holotype of *B. hintzei* (Boyce, 1989, pl. 32, fig. 6), which retains its cuticle. It is not adequate as a specific character, as it depends on preservation. The degree of dorsal expression of segmentation in the pygidium varies in general and is at its least in effaced *B. gibberula*, which shows only three faint axial segments on the internal mould. *B. hintzei* is regarded here as a junior subjective synonym of *B. conica*. Closely similar also is *B. onomeris* Loch, 2007, a species ranging from the *B. stitti* to *S. caudata* biozones of the Kindblade Formation, Oklahoma, and thus broadly equivalent in age to the Nordporten Member. On the cranidium, the palpebral lobe of this species is possibly further away from the glabella at its anterior end than in *B. conica*, but the presence of two pairs of pleural furrows on the pygidial pleural fields is perhaps the only convincing difference. The stratigraphically earlier *B. rhochmotis* Loch, 2007, is convincingly distinguished from *B. conica* in having deeper dorsal furrows generally, longer pre-glabellar field, retained librigenal border furrow and strongly furrowed pygidial pleural fields. It is transitional with *Peltabellia* in several features.

Genus *Ceratopeltis* Poulsen, 1937

Type species. – *Ceratopeltis latilimbata* Poulsen, 1937, Cape Weber Formation, East Greenland.

Discussion. – The type species was known only from an imperfectly prepared pygidium and hence placed among the *incertae sedis* in the 1959 Trilobite *Treatise*, but Fortey & Peel (1983) recognised the genus from better-preserved material from Greenland and associated a cranidium and free cheek. Subsequently, Adrain & Westrop (2005) added an additional species from Ellesmere Island and reinterpreted previous assignments. *Ceratopeltis* is a distinctive genus with two pairs of marginal pygidial spines, presumably a derived member of Bathyurellinae.

Ceratopeltis cf. *C. batchensis* Adrain & Westrop, 2005

Figure 20Q–R

Material. – Incomplete pygidium, PMO 223.228.

Stratigraphic range. – Topmost Bassisletta Member, top of *Chapmanopyge* fauna.

Discussion. – A single pygidium does not provide the basis for a description. However, what there is of it is well preserved, and the stout anterior marginal spine can be seen as well as a sculpture of dorsal terrace ridges. Incomplete though it is, the fragment is identical to specimens figured by Adrain & Westrop (2005). Hence, a comparative determination is made.

Genus *Chapmanopyge* n. gen.

Remarks. – pro *Chapmania* Loch, 2007 (pre-occupied). *Chapmania* has been used as a generic name on several occasions predating Loch, first by Silvesti and Prever (1904) for a member of cestoid worms (*Chapmania*); *Chapmanopyge* is introduced here as a replacement name.

Type species. – *Chapmania oklahomensis* Loch, 2007, original designation.

Remarks. – *Chapmania* (now *Chapmanopyge*) was introduced by Loch for four species from the southern USA, to which he added *Licnocephala sminue* Fortey & Peel, 1990, from Greenland. The genus is clearly allied to bathyurellines such as *Punka* and *Licnocephala*. The species included in *Chapmanopyge* all have a broad pygidium, often with feeble indications of furrows on the wide, flat border, which is bounded at its inner edge by a distinct furrow marking an abrupt change in slope. The cephalic borders are very variably developed in width in the species described by Loch, though a steep break in slope from the pre-glabellar field is typical. Cephalic convexity (sag.) is similar to that of *Bathyurellus*. The glabella may be sub-rectangular in outline or come to a point anteromedially. The

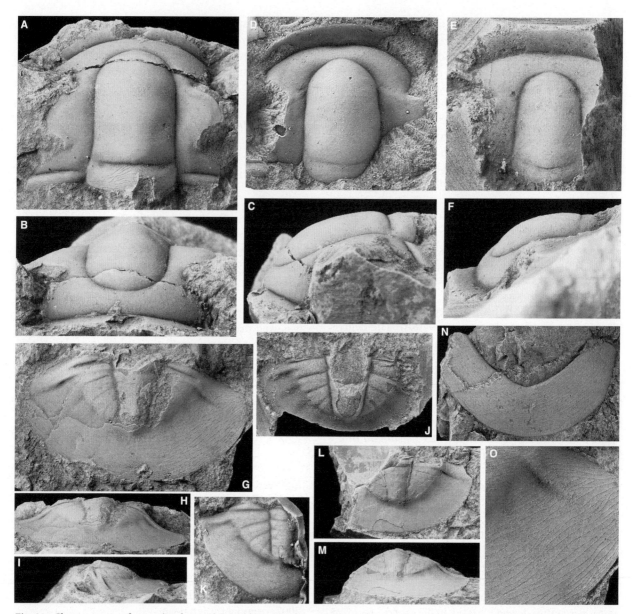

Fig. 21. Chapmanopyge cf. sp. 1 (Loch, 2007). Bassisletta Member, top 34 m (216–250 m from base), late *Chapmanopyge* fauna. **A–C,** Cranidium PMO 208.242 in dorsal, anterior and lateral views (x5), topmost 2 m of Bassisletta Member. **D,** Cast of cranidium CAMSM X 50188.60 in dorsal view (x5), possibly basal Nordporten Member. **E–F,** Cranidium PMO 208.305 in dorsal and lateral views (x6), horizon as in (**A**) (wider border). **G–I,** Large pygidium PMO 208.235 in dorsal (x2), posterior and lateral views (x1.5), and (**O**) enlargement (x4) of **G** showing sculpture, top 2 m of Bassisletta Member. **J,** Incomplete small pygidium, more strongly furrowed, PMO 221.271 in dorsal view (x4), 34 m below top of Bassisletta Member. **K,** Latex cast of pygidium close to Loch's material PMO 221.255 in dorsal view (x3), horizon as **A**. **L–M,** Very large incomplete pygidium PMO 208.238 in dorsal and posterior views (x1), same bed. **N,** Pygidial doublure PMO 208.281 in dorsal view (x1.5), same bed.

problem of discrimination of *Chapmanopyge* from *Licnocephala* resides in the imperfect knowledge of the type species of the latter (see p. 21). However, even given the small size of type specimens of *L. bicornuta* Ross, 1951, very low cephalic convexity and the absence of an incised edge to the border are likely real features that separate it from *Chapmanopyge*. If the pygidium of *Licnocephala bicornuta* is correctly assigned, it, too, lacks a clearly demarcated

border. So one might make a case for the more convex species selected by Loch as an early clade of bathyurellines. However, the sub-rectangular glabella of some species could be a plesiomorphic character, shared as it is with some species of the 'stem' bathyurid *Peltabellia*, and given the several generic names available for generally similar bathyurids, the concepts need to be evaluated in a full review.

Chapmanopyge cf. sp. 1 (Loch, 2007)

Figure 21

cf. 2007 *Chapmania* sp. 1 Loch, pl. 23, fig. 10.

Material. – Cranidia, PMO 208.233, 208.242, 208.305, CAMSM X 50188.60; pygidia, PMO 208.235, PMO 208.280–1, 208.287, 221.255; growth stages, PMO 221.269, 221.271.

Stratigraphic range. – Top 34 m of Bassisletta Member, *Chapmanopyge* fauna and probably basal bed of Nordporten Member.

Description. – Cuticle thin. Cranidium was not associated by Loch (2007), but the one figured here seems a plausible candidate from the same bed as the best pygidia, of similar moderate convexity (sag.) to others attributed to the genus. The glabella is longer than that of most other species of the genus, length 1.4 times maximum width at occipital ring and tapering regularly anteriorly until rounded, or at least scarcely acuminate glabellar front. Occipital furrow narrow, fading laterally. Pre-glabellar field sloping, in dorsal view, <20% glabellar length. Anterior sections of facial sutures diverge at about 35° to sag. line before curving adaxially distally. Palpebral lobes not close to glabella but not fully preserved. Flat anterior border defined by distinct, narrow furrow border as wide (sag.) as pre-glabellar field in dorsal view.

Pygidium very transverse, 0.5–0.6 times as long as wide, and the border 45% to 50% as long as the axis (sag.) and slightly wider laterally. Axis tapers gently to somewhat truncate tip; width of the axis anteriorly (imperfectly preserved) is about the same as the adjacent pleural field. Only one ring is clearly defined. Four pairs of pleural furrows not extending noticeably on to border; the anterior pair longest and deepest, posteriorly progressively fainter and more sloping backwards. Facet is backed by short section of deep interpleural furrow. There is a very weak indication of a second interpleural furrow on the proximal part of border. Edge of pleural field is marked by steep change in slope and also by paradoublural line. The smaller pygidium (Fig. 21K) shows deeper marginal furrow at edge of border and clearer indications of the lateral parts of the posterior three axial ring furrows. Border slopes gently and is slightly dished anterolaterally. Posterior pygidial outline a shallow arc of a circle. Surface sculpture of very weak terrace ridges densely packed and sub-parallel to margin, except on facet where they are stronger and oblique.

Discussion. – Described species of *Chapmanopyge* have characteristic pygidia. The species described from the Kindblade Formation by Loch (2007) include one species with a tapering glabella very much like that described above, namely *C. taylori* (Loch), but its pygidium is relatively much longer than that of our species. *C. carteri* (Loch) has a considerably narrower pygidial border. Two species have pygidia much like the species from the Kirtonryggen Formation: *C. oklahomensis* (Loch) and '*Chapmania* sp. 1'. *C. oklahomensis* has a quite different cranidium from that we have associated with our species, with a large glabella that does not taper anteriorly. '*Chapmania* sp. 1' (Loch 2007, pl. 23, fig. 10) compares in all details with our specimen on Figure 21K, even to the extent of having a weak proximal part of the second pleural furrow, but it is without an associated cranidium. The pygidial axis is slightly more slender than in the pygidium of *C. oklahomensis*, which is consistent with a narrow (tr.) glabella. Our larger pygidia are comparatively more effaced (Fig. 21G). *Chapmanopyge sminue* (Fortey & Peel, 1990) from the Poulsen Cliff Formation, Greenland, was originally assigned to *Licnocephala*, but Loch reassigned it to *Chapmania* (i.e. *Chapmanopyge*), correctly on the criteria listed above. It also has a similar pygidium, but with a very convex and longer axis in relation to the border, and its glabella is almost rectangular. Despite having a cranidium in Spitsbergen, lack of other cephalic parts, and imperfect preservation, oblige us to retain Loch's (2007) open nomenclature rather than formally erect a new taxon. But we do regard the Spitsbergen species and the Oklahoma one as very likely one and the same.

Chapmanopyge '*amplimarginiata*' (Billings, 1859)

Figure 22

1859 *Bathyurus amplimarginiatus* Billings, p. 365, fig 12.

1865 *Bathyurus amplimarginiatus* Billings; Billings, p. 353, fig. 341a.

1938 *Bathyurellus amplimarginiatus* Billings; Twenhofel, p. 71, pl. 10, fig. 13.

2011 *Peltabellia amplimarginiata* (Billings); Shaw and Bolton, p. 413, fig. 4.2, 4.4, *non* fig. 4.3, 4.5, 4.6.

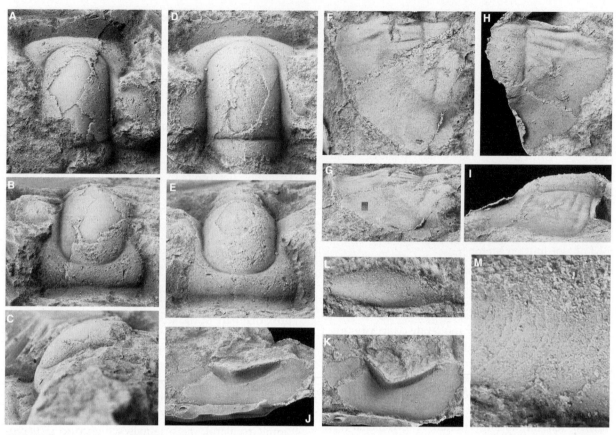

Fig. 22. Chapmanopyge 'amplimarginiata' (Billings, 1865). Bassisletta Member, all 114 m from base, early *Chapmanopyge* fauna. **A–C,** Incomplete cranidium PMO 223.194 in dorsal, anterior and lateral views (x4). **D–E,** Cranidium PMO 223.195 in dorsal and anterior views (x6). **F–G,** Latex cast from incomplete pygidium PMO 223.198 in dorsal and posterior views (x2). **H–I,** Latex cast from another incomplete pygidium PMO 223.196 in dorsal and lateral views (x5). **J–K,** Pygidial doublure PMO 223.200 in posterior and dorsal views (x2) showing very sharply reflexed inner edge. **L,** Poorly preserved incomplete free cheek PMO 223.199 in lateral view (x3) with (**M**) enlargement (x12) showing terrace lines on genal field.

Material. – Cranidia, PMO 223.194–195; free cheeks, PMO 223.199; pygidia, PMO 223.196, 223.198, 223,200.

Stratigraphic range. – Bassisletta Member 104 m from base. Basal occurrence of the *Chapmanopyge* fauna.

Discussion. – Another species we attribute to *Chapmanopyge* occurs about 70 m below the range of *Chapmanopyge* sp. 1 and is clearly different. It is of stratigraphic importance as the only trilobite found in the least fossiliferous part of the Bassisletta Member. Pygidia are known from incomplete examples, but display a much more deeply curved posterior outline than that of *C.* sp. 1, although the border is as wide, or wider. Cranidia from the same bed as the pygidia have a more rectangular outline and wider (tr.) anterior fixed cheeks, and the palpebral lobes are closer to the glabella. The anterior cranidial border is of comparable width (sag.). The little-known species *Bathyurus amplimarginiatus* was described by

Billings from the Romaine Formation of the Mingan Islands, Quebec, and recently redescribed by Shaw & Bolton (2011). The material is also incomplete, but the pygidium shows a similar outline to the material under discussion, and with a comparably wide border, with interpleural furrows somewhat more prominent on the adaxial part of the pygidial pleural field than in *Chapmanopyge* sp. 1 (Loch, 2007). Of the other species described by Loch (2007) from Oklahoma, *C. taylori* has a pygidium with a similar outline, and the cranidium is very similar to the Spitsbergen form, but the pygidial border is not so wide, and the more inflated proximal part of the pygidial pleural field is more clearly defined along its outer edge. Shaw & Bolton (2011) placed *amplimarginiatus* in *Peltabellia*, and it is true that the inner edge of the pygidial border is not as defined as in other *Chapmanopyge* species. However, we do have a doublure which presumably belongs here that shows an extremely sharply upturned inner edge, with a groove (i.e. ridge ventrally) at its base. This structure is consistent with the sharply bounded edge of the border in other *Chapmanopyge* spp. Cranidial fragments associated with *C. amplimarginiata* by Shaw and Bolton seem more likely to belong to *Jeffersonia*. The comparative material is not entirely adequate, and neither is our material, but we would not feel justified to use any other than the Billings name *pro tem,* but we use it with caution. It is interesting to note that the somewhat similar species from Oklahoma is the oldest *Chapmania* recorded there.

Chapmanopyge cf. *C. sanddoelaensis* (Adrain & Westrop, 2005)

Figure 23I–M

Material. – Pygidia, PMO 208.149, 208.151.

Stratigraphic range. – Top Bassisletta Member, *Chapmanopyge* fauna.

Discussion. – These pygidia are distinctive, but we have not been able to recognise associated cephalic sclerites, and hence, open nomenclature has to be used for them. The pygidia are about twice as wide as long, and the broad border maintains a uniform width along its length. The axis is relatively effaced, short and wide, truncate posteriorly, and four axial rings are visible. The narrowly triangular pleural fields are very weakly marked by pleural furrows, with the third pair hardly visible. The border is clearly convex, with a particular downward slope surrounding the posterior margin, as opposed to the flat, or even slightly concave, profile of other *Chapmanopyge* spp. However, a species described from Ellesmere Island as *Licnocephala sanddoelaensis* by Adrain & Westrop (2005) has a deeply defined boundary between pygidial pleural fields and border, a character regarded herein as a way of separating *Chapmanopyge* from *Licnocephala*. Like our material, this species also has a convex border. Although most pygidia illustrated by Adrain & Westrop (2005) have

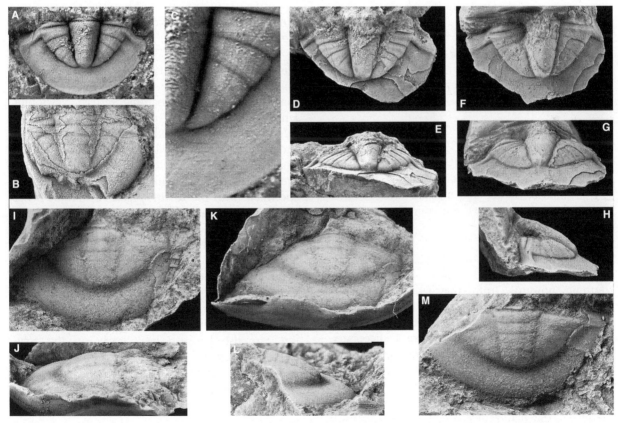

Fig. 23. **A, C–H,** *Chapmanopyge* n. sp. A. Nordporten Member, known from 105 m *P. groenlandicus* fauna. **A, C,** Pygidium CAMSM X 50188.28 in dorsal view (×6) and (**C**) enlargement showing the sculpture (×18), probably low *P. nero* fauna exact horizon uncertain. **D–E,** Pygidium CAMSM X 50188.46 in dorsal and posterior views (×2), probably low *P. nero* fauna. **F–H,** Pygidium PMO 223.248 in dorsal, posterior and lateral views (×2), 105 m. **B,** *Chapmanopyge* cf. n. sp. A. Pygidium CAMSM X 50188.41 in dorsal view (×2) *P. nero* fauna, ca. 180 m. **I–M,** *Chapmanopyge* cf. *C. sanddoelaensis* (Adrain & Westrop, 2005). Top 2 m of Bassisletta Member, younger *Chapmanopyge* fauna. **I–K,** Latex cast of pygidium PMO 208.151 in dorsal, lateral and posterior views (×4). **L–M,** Internal mould of pygidium PMO 208.149 in lateral and dorsal views (×4).

a more deeply curved posterior profile and well-defined pleural furrows, two of their illustrated pygidia (their figs 7.2 and 7.9) are not very different from the specimens from the Kirtonryggen Formation in this regard, and we did consider that the species under discussion might be conspecific. However, all the pygidia from Ellesmere Island have narrower and convex (tr.) pygidial axes that are not truncated where they meet the border, but come to a point. Assignment to *Chapmanopyge* rather than *Licnocephala* does seem appropriate given the transverse pygidium with sharply defined border. However, the convex border is different from most species of either genus. We also considered the possibility that the Kirtonryggen species could belong to *Randaynia*, since the pygidium of *R. leatherburyi* Loch, 2007, is generally similar, notably in its clear axial segmentation combined with weak pleural furrows, low convexity of the axis, which is posteriorly truncate, and convex pygidial border. On balance, we considered that the resemblance to *C. sanddoelaensis* was more compelling, but cephalic sclerites are needed properly to clarify this species.

Chapmanopyge n. sp. A

Figure 23A, C–H

Material. – Pygidia, PMO 223.248, CAMSM X 50188.28, X 50188.46.

Stratigraphic range. – Nordporten member, *P. groenlandicus* fauna 105 m, and extending to *P. nero* fauna (lower part), but collections not precisely located.

Discussion. – The few pygidia discussed here are of interest as showing that *Chapmanopyge* species continue upwards from the Tulean into later Ibexian strata that are Blackhillsian in age. We have not certainly identified cephalic sclerites for this taxon, and open nomenclature is necessary. The pygidia have a relatively deep profile like that of *C. taylori* Loch, 2007, but lack clear traces of the interpleural furrows seen on that species, with narrower but deep pleural furrows, stopping sharply at the flat, gently declined border, and a shorter, relatively wider axis, with four weak rings. Surface sculpture terrace ridges on centre of axis, very sparsely on border. There may prove to be yet more species, because one of the pygidia (referred to provisionally as cf. sp. A on

Fig. 23B) has a distinctly more convex border (the doublure is reflexed horizontally), the inner edge of which is less sharply defined than in other specimens, and a wider axis. These interesting taxa would clearly benefit from more collections.

Genus *Grinnellaspis* Poulsen, 1946

Type species. – *Actinopeltis feildeni* Poulsen 1946, Ellesmere Island.

Grinnellaspis newfoundlandensis Boyce, 1989

Figure 24A–C

1989 *Grinnellaspis? newfoundlandensis* sp. nov. Boyce, pp. 58–59, pl. 33, figs 1–6.

Material. – Pygidium, PMO 223.126.

Stratigraphic range. – Nordporten Member 187 m from base, *P. nero* fauna.

Discussion. – Boyce (1989) described *G. newfoundlandensis* from the Barbace Cove Member of the Boat Harbour Formation, western Newfoundland. The pygidium is unique among bathyurids in having a posteromedian slit behind the axis. Indeed, there is no trilobite known to us with exactly this structure, although median embayments are not uncommon in several families. The specimen from the Kirtonryggen Formation is better preserved than Boyce's type pygidium, but like it shows four more or less radially disposed pleural furrows crossing the broad, gently sloping pygidial border and narrowly triangular convex pleural fields cut by three narrow, deep furrows. The sculpture of undulating fine terrace ridges running approximately transversely across the pygidium is beautifully shown on our specimen, but can also be seen less clearly on the Newfoundland material. Boyce had fragmentary cephalic material, which we lack, but in view of the remarkable pygidium, there is no reason to doubt that the same species is represented in Spitsbergen.

Genus *Harlandaspis* n. gen.

Type species. – *Harlandaspis elongata* n. sp. (below).

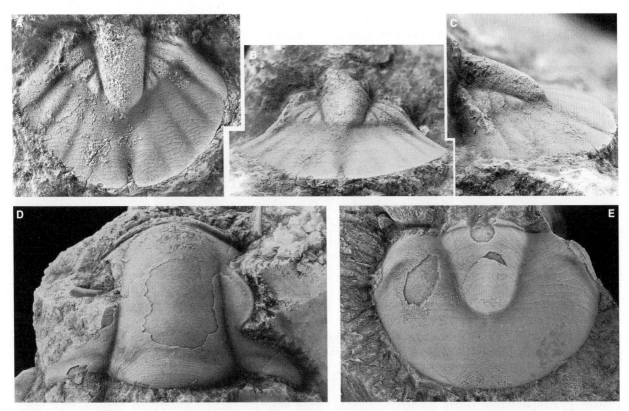

Fig. 24. **A–C**, *Grinnellaspis newfoundlandensis* Boyce, 1989. Nordporten Member, *P. nero* fauna, 187 m, pygidium PMO 223.126 in dorsal, posterior and lateral views (x5). **D–E**, *Harlandaspis elongata* n. gen., n. sp. Enlargements from next figure of type specimens to show surface sculpture. Nordporten Member, *P. groenlandicus* fauna, 105 m. **D**, Holotype cranidium PMO 223.175 in dorsal view (x4). **E**, Pygidium PMO 208.132 in dorsal view (x5).

Diagnosis. – Large Bathyuridae with Bertillon type sculpture. Glabella well defined, gently tapering and unusually elongate (sag.), with weakly incised occipital furrow. Pre-glabellar field reduced or absent medially. Narrow, convex, brim-like cephalic anterior border. Eyes (and palpebral lobes) of moderate size and gently curved. Wide (tr.) free cheeks lacking genal spine, genal field more or less bisected by weak furrow; doublure with vincular notch. Convex pygidium very much like that of *Benthamaspis*, with axis tapering to half pygidial length, lacking border but with strong facets, inner edge of doublure curved sharply dorsally.

Species included. – *Harlandaspis elongata* n. sp., *Bathyurellus inflata* Loch, 2007.

Etymology. – The genus is named for W. B. (Brian) Harland, who contributed much to the geology of Spitsbergen, and introduced the first author to the island.

Discussion. – The genus is erected to include distinctive bathyurids with elongate, weakly convex (sag.) glabella, uptilted and narrow cephalic border, reduced pre-glabellar field and wide (tr.) free cheeks without genal spines. The pygidial morphology and sculpture are very much like that of *Benthamaspis*, but the cephalic morphology could scarcely be more different. All *Benthamaspis* species are small, have a very short, often sub-rectangular glabella, and the cephalic border is reduced to an exceedingly thin rim; the eyes are invariably large, and the free cheek narrow (tr.). It would not be possible to accommodate the species described below as *Harlandaspis elongata* into an expanded concept of *Benthamaspis*. However, the upturned cephalic border that continues on to the free cheek with a forwardly facing strip of doublure is similar in both genera and coupled with nearly identical pygidia indicates either descent from a common ancestor or a striking degree of convergence. The surface sculpture on *Harlandaspis elongata* is similar to the Bertillon that densely covers the cuticle of *Benthamaspis* species, although more spaced out. The species described by Loch (2007) as *Bathyurellus inflata* from Oklahoma has a pygidium almost identical to that assigned to *Harlandaspis elongata*, but its cranidium retains a pre-glabellar field. However, small examples of cranidia of *H. elongata* (Fig. 25V) *do* have a longer (sag.)

pre-glabellar field than many large specimens, on some of which (Fig. 25O) it becomes reduced to nothing. The small cranidium of *H. elongata* is otherwise very similar to that of Loch's species, which is considered to represent a second species of *Harlandaspis*. It is likely that the Oklahoma species

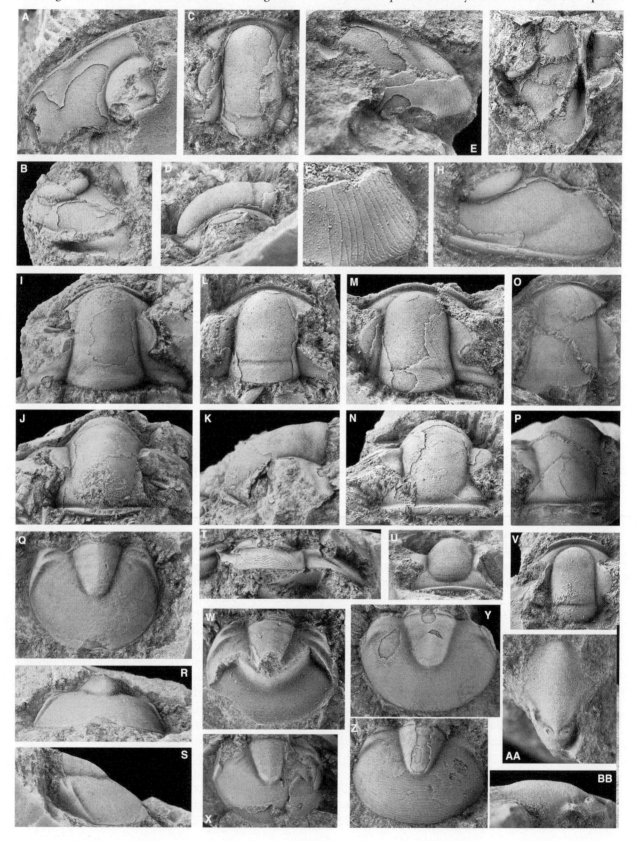

is more plesiomorphic, retaining a pre-glabellar field from the common ancestor of both *Harlandaspis* and *Benthamaspis,* and its general similarity to *Bathyurellus* may be significant in this regard.

Harlandaspis elongata n. sp.

Figures 24D–E, 25

Holotype. – Cranidium, PMO 223.175 (Figs 24D, 25I–K); Nordporten Member, 105 m.

Material. – Cranidia, PMO 208.088, 208.103, 208.113, 208.116–117, 223.130, 223.151, 223.154, 223.157; free cheeks, PMO 208.115, 208.144 (1), 223.122–123, 223.152, 223.190, 223.193; pygidia, PMO 208.107, 208.130, 208.132, 208.133, 208.134, 208.136, 208.174; hypostome, PMO 221.243; thoracic segment, PMO 223.190.

Stratigraphic range. – Lower part of Nordporten Member, 105–130 m from base (*Petigurus groenlandicus* fauna).

Diagnosis. – *Harlandaspis* lacking pre-glabellar field at large size; surface sculpture of widely spaced Bertillon pattern; pygidial axis sub-conical.

Etymology. – 'Elongate' referring to glabella.

Description. – Collections from 105-m horizon allow confident association of sclerites. The cuticle is thin, especially when compared with other large bathyurids like *Petigurus.* Exfoliation is common, and internal moulds do not record the characteristic surface sculpture. Species grows large, with fragmentary specimens suggesting cranidial length of up to 4 cm. Cranidium shows rather low convexity (sag.) compared with many bathyurids, glabella being horizontal in its posterior part and curving down rather gently and evenly forwards, and transversely not greatly vaulted. Glabella long and gently tapering forwards, from more than two-thirds to almost twice

as long (sag.) as width at mid-length, the longest also being the largest, frontal lobe having an evenly gentle arcuate outline in dorsal view. Axial furrows are well defined and often slightly pinched inwards along a line near the posterior end of the palpebral lobes. Pre-glabellar furrows shallower on large cranidia. Occipital ring defined by a shallow occipital furrow, widest medially, tending to effacement altogether on largest cranidia. No sign of glabellar furrows. Pre-glabellar field is very short (sag.) to absent medially and is reduced progressively on large cranidia. Smallest cranidium (Fig. 25U) shows a distinct, though narrow pre-glabellar field across the midline. Palpebral lobes close to glabella 0.3–0.4 length (exsag.) of cranidium (sag.), medially positioned, with gently curved outline, less than a semicircle (many bathyurids are more strongly curved). Anterior branches of facial sutures moderately divergent in front of the eyes, with straight course to cut border at acute angle of about 60–70° as seen in anterior view. Posterior suture branches diverge more strongly to near transverse before an elbow beyond the palpebral lobe whence they curve backwards to cut the posterior margin at a low acute angle. Transverse width of posterior fixed cheek is quite narrow for a bathyurelline, no more than two-thirds transverse width of occipital ring. A distinct, narrow border furrow outlines the anterior cranidial border, which is convex, but with a slightly sharp edge separating the dorsal part from a sloping, anteriorly facing or recurved anterior wall. Posterior limb almost bisected by shallow posterior border furrow. Spaced terrace ridges, often fine raised lines, curve gently forwards over occipital ring, and more prominently so over the glabella; absent on palpebral lobe.

Free cheek wide (exsag. about 60% of transverse width in plan) with rounded genal angle. Steep narrow rim continues from cranidium, but becomes less prominent laterally and does not quite reach genal angle as the furrow defining it fades. Genal field is approximately bisected by a furrow that curves inwards–backwards from the end of the deep part of the lateral border furrow. We consider this furrow to be a continuation of the border furrow, that is, the

---

Fig. 25. Harlandaspis elongata n. gen., n. sp. Nordporten Member, *Petigurus groenlandicus* fauna, 105 m unless stated. **A**, Free cheek PMO 208.144 (1) in dorsal view (x4). **B**, Free cheek with doublure prepared PMO 223.193 in lateral view (x3). **C–D**, Cranidium with long glabella PMO 223.130 in dorsal and lateral views (x4). **E, F**, Free cheek PMO 223.122 in lateral view (x4) and (**F**) enlargement (x8) showing sculpture on rounded genal angle. **G**, Free cheek with doublure PMO 223.152 in dorsal view (x3) and vincular notch indicated. **H**, Free cheek PMO 208.115 in anterolateral view (x4), illustrating no eye socle. **I–K**, Holotype cranidium PMO 223.175 in dorsal, anterior and lateral views (x2.5), (see Fig. 24D). **L**, Cranidium PMO 223.151 in dorsal view (x4). **M–N**, Cranidium PMO 223.157 in dorsal and anterior views (x5). **O–P**, Large cranidium showing longer glabella PMO 208.103 in dorsal and anterior views (x2). **Q–S**, Pygidium PMO 208.174 in dorsal, posterior and lateral views (x2). **T**, Thoracic segment PMO 223.190 in dorsal view (x3). **U–V**, Small cranidium PMO 223.154 in anterior and dorsal views (x5). **W**, Pygidium prepared to reveal doublure PMO 208.133 in dorsal view (x3). **X**, Pygidium PMO 208.134 in dorsal view (x2). **Y**, Well-preserved pygidium PMO 208.132 in dorsal view (x3) (see Fig. 24E). **Z**, Pygidium PMO 208.136 in dorsal view (x5) showing similarity to that of *Benthamaspis* at this size. **AA–BB**, Hypostome PMO 221.243 in dorsal and anterior views (x5), 130 m.

true border lies outside it and expands in width posteriorly and laterally and is gently inflated. This is
demonstrated by the course of the doublure that lies
under that part of the cheek and narrows anteriorly
in line with the border furrow. As it does so, it is also
bent into a deep tube around the median anterior of
the cephalon, in conjunction with the anterior
cephalic border. The lower surface of the doublure
carries terrace lines running sub-parallel with the
cephalic margin. Its inner edge has a small vincular
notch in its outline at about the point it achieves
maximum width from the front border. Eye lobe is
rather low and gently curved with impalpably fine
holochroal lenses and sits directly on the genal field
without separately inflated eye socle. Cephalic style
of sculpture is present on the posterolateral border,
but sparser on posterior part of genal field in some
specimens.

We associate a hypostome with this species
(Fig. 25AA) because it has an unusually long attenu-
ated anterior lobe of the middle body and seems
appropriate for a trilobite with an exceptionally long
glabella. If we correctly interpret the hypostome for
Uromystrum herein, there seems no other suitable
trilobite known to us in the fauna for this
hypostome, and an association with *Harlandaspis
elongata* is reasonable. A pair of maculae lie close to
the rounded posterior margin, and the posterior
lobe of the middle body is very short (sag., exsag.).
The fundamental design is not unlike that of the
hypostome associated here with *Jeffersonia* drawn
out, as it were, into a longer version. It is noted that
there is a long pair of anterior wings dorsally directed that is consistent with attachment to the inner
edge of the tube-like ventral structure described in
the previous paragraph.

Pygidium 1.15–1.4 times wider than long in dorsal
view, convex (sag., tr.), with a rather evenly convex
curvature in profile down to the posterior margin
from front to back. Axis convex, tapering (axial furrows enclosing an angle of about 30°), sub-conical
in outline, extending to half, or slightly more pygidial length in dorsal view, and occupying 40% of

maximum pygidial width at its anterior margin.
There is a suggestion of a post-axial ridge on several
specimens. Apart from a narrow (sag.) half-ring,
dorsal furrows are faint on the axis, but internal
moulds show three sub-equal rings extending over
more than half axial length, the third one faint. Similarly, pleural fields are not furrowed dorsally after
the first half rib, but internal moulds show up to
four faint pleural and/or interpleural furrows close
to the axis, not extending on the pleural field underlain by wide doublure. Border lacking. Anterior
pygidial segment is quite well defined, with the pleural shoulders making a ridge on the flanks of the
pygidium and a broad, almost vertical facet sloping
slightly outwards and strongly backwards. Many
specimens break off around the robust doublure,
which is horizontal over much of its length, but
turned up sharply around the axial tip and adjacent
pleural fields. Dorsal surface carries fine, sparsely
anastomosing terrace ridges that are more transverse
than the almost semicircular posterior pygidial margin. Similar ridges are present on the ventral surface
of the doublure.

Discussion. – The surprising aspect of this species is
the strong similarity of pygidia to those of *Benthamaspis*, although they are much larger. It is noted that
the free cheeks of *Benthamaspis* species show a previously inexplicable ridge below the eye. This feature
can be explained with reference to the border on the
cheek of *Harlandaspis*, if we have the homology correct. The true border remains on *Benthamaspis*, occupying virtually all the genal area on the free cheek.
The intermediate stage is illustrated by the species
B. rhochmotis Loch (2007, pl. 21, fig. 7) on which a
moderately wide lateral border is still present. The
inner part of the cheek, as seen as a wide area (tr.) on
Harlandaspis, has become reduced just to the proximal part of the border furrow alone in advanced
Benthamaspis. If correctly interpreted, this feature
provides further evidence of a sister group relationship between *Harlandaspis* and *Benthamaspis*. *Harlandaspis elongata* obviously differs from the only

Fig. 26. **B–I**, *Licnocephala brevicauda* (Poulsen, 1937). Nordporten Member, probably entirely *P. groenlandicus* fauna. **B–C**, Partial cranidium CAMSM X 50188.37 in dorsal and lateral views (x4), unspecified horizon. **D–E**, Free cheek PMO 208.138 in dorsal view (x1.5),
105 m, with enlargement (**E**) showing sculpture on genal area (x6). **F–H**, Pygidium CAMSM X 50188.35. in dorsal, posterior and lateral
views (x5), exact horizon unknown. **I**, Holotype, original of Poulsen 1937, pl. 7, fig. 9 (x2). **A**, *Licnocephala* aff. *brevicauda* (Poulsen),
P. nero fauna, free cheek PMO 208.196 in dorsal view (x2), 190 m, differing from other specimens in uninterrupted lateral border. This
free cheek could also belong to *Licnocephala* n. sp. A, but is from a different stratigraphic horizon to the pygidia. **J–L**, *Licnocephala* n. sp.
A. Nordporten Member, **H** collection, low in *P. groenlandicus* fauna (exact horizon uncertain). **J**, Pygidium PMO 208.252 in dorsal view
(x2). **K–L**, Pygidium PMO 221.235 in dorsal and lateral views (x2). **M, O**, *Licnocephala* n. sp. B. **M**, Latex cast of pygidium CAMSM X
50188.58 in dorsal view (x2), exact horizon unknown. **O**, Pygidium PMO 208.254 in dorsal view (x2.5), **H** collection, low in *P. groenlandicus* fauna, exact horizon unknown. **N**, *Licnocephala* aff. n. sp. B. Pygidium PMO 208.079 in dorsal view (x3), 106 m. **R–U**, *Licnocephala*? n. sp. C. Topmost 2 m Bassisletta Member. **R**, Pygidium PMO 208.295 in dorsal view (x3). **S–U**, Pygidium PMO 208.275 in lateral,
dorsal and posterior views (x5). **P–Q**, *Licnocephala* sp. Topmost Bassisletta Member. **P**, Free cheek PMO 221.254 in dorsal view (x1.5).
Q, Cranidium PMO 223.238 in dorsal view (x4).

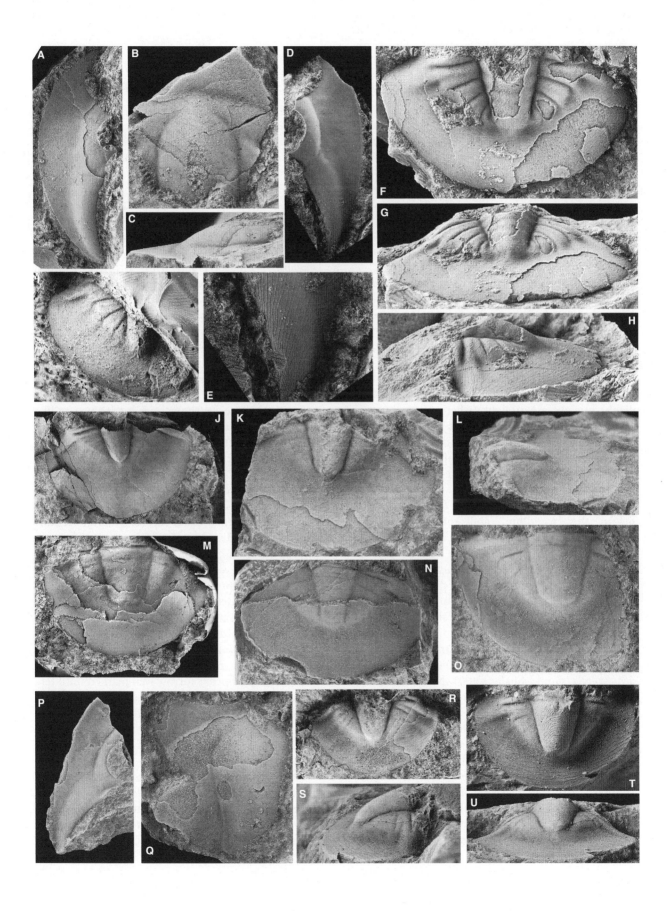

other species assigned to the genus, *H. inflatus* (Loch, 2007), in lacking a pre-glabellar field and in the development of the typical Bertillon sculpture. The pygidia of the two species are similar. The cranidium of *H. inflatus* is much more like that of the generalised bathyurelline, resembling the immature cranidium of *Bathyurellus diclementsae* illustrated here as Figure 19C, for example. This provides further evidence of an ultimate common ancestor for *Bathyurellus* and *Benthamaspis*, *Harlandaspis* being closer to the latter.

Genus *Licnocephala* Ross, 1951

Type species. – *Licnocephala bicornuta* Ross, 1951; Garden City Formation, Idaho, original designation.

Remarks. – As noted previously, the type species of *Licnocephala* is not well known, and the type material comprises relatively small sclerites. The genus is employed here for bathyurellines with thin cuticle that have very low convexity (sag.) on both cephalon and pygidium, often combined with a wide, flat anterior cephalic border. The pygidium retains the primitive condition wherein the pleural and interpleural furrows are straight (where seen), but the outer margin of the pleural field is not defined by a sharp furrow or steep slope, as it is in *Chapmanopyge* and *Punka*, but only by a comparatively gentle change in slope. These features are recognisable on the type species; however, Ross (1951) only cautiously assigned a pygidium to *L. bicornuta*. Nonetheless, this pygidium, and one placed with *L. cavigladius* Hintze, 1953, do resemble the pygidia described below from Spitsbergen and differ from contemporary species of *Chapmanopyge* in the definition of the border. Ross (1953) provided further information on *L. cavigladius,* and one of his pygidia (Ross 1953, pl. 64, fig. 25) shows short sections of the interpleural furrow passing on to the border, without, however, the sigmoidal form we regard as typical of *Punka*. Redescription of the type species of *Licnocephala* should finally settle whether *Chapmanopyge*, *Punka* and *Licnocephala* can be maintained as separate genera. All three do seem to have independent, contemporary stratigraphically significant species, and it would be difficult to place such different species as *L.* n. sp. A (below) and *Punka flabelliformis* in the same genus. However, the pygidia assembled here into *Licnocephala* show a range in morphology, and it does seem likely that more than one clade will be identified in the future. Unfortunately, our material is not adequate formally to characterise new taxa.

Licnocephala brevicauda (Poulsen, 1937)

Figure 26B–I

1937 *Niobe brevicauda* Poulsen, p. 56, pl. 7, figs 9, 10.

Material. – Incomplete cranidium, CAMSM X 50188.37; free cheek, PMO 208.138; pygidium, CAMSM X 50188.35.

Stratigraphic range. – Nordporten Member, collected from 105 m horizon, and from several isolated localities, with *Petigurus groenlandicus* fauna.

Diagnosis. – *Licnocephala* with furrowed pygidial pleural fields, first three pairs pleural furrows hardly turning posteriorly distally, but fourth pygidial pleural furrow turns back to define a narrow triangular posterior pleural lobe. Surface sculpture of undulating terrace ridges on wide border and pleural fields.

Description. – Species of medium size, with noticeably thin cuticle. The incomplete cranidium associated with this species shows very low convexity like that of the pygidium. Glabella relatively well defined, parallel sided, but with low transverse profile, coming to an anteriorly acuminate termination where its definition is less clear. Pre-glabellar field relatively short. Shallow border furrow arching forwards around mid-line separates wide, flat, anterior border from anterior parts of fixed cheeks. Anterior sections of facial sutures divergent. Eye ridges directed posteriorly to front end of palpebral lobes, which are not preserved (they were presumably moderately large). Free cheek of appropriate morphology has low convexity, wide border and border furrow of appropriate depth to continue from cranidium. At about half its length, the border narrows markedly where interrupted by a posteriorly inflated portion of the genal field, continuing on into broadly triangular genal spine. Posterior border furrow well marked until terminated by the same inflated area. The latter is set off from the rest of the genal field by a shallow furrow that probably represents the true course of the lateral border furrow, that is, the inflated area is really part of the border.

Flattish pygidium 1.8–1.9 times as wide as long, with the border wider than the pleural lobe. Weakly convex axis occupies a quarter maximum pygidial width at anterior margin and extends a little more than half its length (sag.), gently tapering, with four rings of which as few as two may be easily visible. A posteromedian swelling is visible at the posterior end of the axis, where the axial furrows shallow around the terminal piece. Pleural lobe defined by change in slope along its outer edge but no furrow. Of the four narrow pleural furrows, the first, leading to the facet, is deepest and most transverse. The remaining three furrows slope progressively steeply backwards, hardly curving posteriorly rearwards at their outer ends along the slope at the edge of the pleural field, and do not continue on to border. Apart from at the adaxial rear side of the facet, where it is marked, the anterior interpleural furrow continues weakly towards the axis; other interpleural furrows are very weak. The posterior pleural furrow is characteristically more steeply curved backwards, to outline an almost drop-like posterior pleural lobe adjacent to the terminal piece of the axis. Facet extends across at least half width of anterior pleural margin. Posterior border slopes downwards gently, may be slightly convex. Surface sculpture of weak terrace ridges running slightly oblique to pygidial margin, and anastomosing occasionally, extending on to pleural fields.

Discussion. – Poulsen (1937) identified two species from pygidia from Greenland in the asaphid genus *Niobe*: *N. groenlandica* and *N. brevicauda*. Both of them are typical bathyurids, although with their morphology convergent upon that of *Niobe* so that Poulsen's judgement is understandable. *Niobe groenlandica* is known from one incomplete pygidium, but this does show similar structure of the pleural fields, with distally curved pleural furrows, to material collected from the lower part of the Nordporten Member of the Kirtonryggen Formation assigned here to *Punka latissima* n. sp. Here we transfer *N. groenlandica* to *Punka*. The dorsal surface of the flat pygidium of '*Niobe*' *brevicauda* (Fig. 26I) has straighter pleural furrows that continue very weakly on to the pygidial border, together with faint interpleural furrows, similar to the pygidium of the species described above. Hence, we use Poulsen's name for our material, in spite of the fragmentary nature of the holotype, and transfer the species to *Licnocephala*. The pygidial doublure attributed to the same species by Poulsen may belong, but could belong equally, to *Punka* spp. The pygidium is also similar to those assigned to species from the Great Basin, *Licnocephala bicornuta* Ross, 1951 (a tentative assignment), and *L.? cavigladius* Hintze, 1953, in its very flat morphology and poorly demarcated pygidial pleural fields. It is rather less similar to three other species from Spitsbergen described below under open nomenclature. The thin cuticle of *Licnocephala* spp. militates against the preservation of fragile cranidia in inshore palaeoenvironments.

Licnocephala n. sp. A

Figure 26J–L

Material. – Pygidia, PMO 208.252, 221.235.

Stratigraphic range. – Nordporten Member, H collection, probably low *Petigurus groenlandicus* fauna.

Diagnosis. – Large species with long pygidium (sag.) having deeply parabolic outline, very narrow pleural fields poorly differentiated from exceedingly wide, flat border and unfurrowed axis well under half pygidial length.

Description. – Large pygidium with thin cuticle 1.4–1.5 times wider than long and very flat. Outline deeply parabolic. Transversely convex axis narrow, about one-quarter pygidial width at anterior edge and less than half pygidial length (40–45%), tapering to posteriorly rounded end, which shows a short, post-axial medially inflated area or ridge. Axial rings faintly defined dorsally. Pleural lobes flat, not sharply differentiated from border, but outer edges converge backwards at a high obtuse angle, and with straight margin. Only anterior pleural furrow deep, straight and nearly transverse, extending to facet. Anterior pygidial segment defined, with interpleural furrow extending on to border behind facet, marked by backwards-sloping ridge. Other pleural furrows obscure. Extremely wide pygidial border gently 'dished' to wide, flat margin. This is the widest border on any bathyurid. Sculpture of weak, fine terrace lines on outer part.

Discussion. – This species is highly distinctive, but known from only one collection originating from the mid–low part of the Nordporten Member, outside the main section, but probably low *P. groenlandicus* Fauna. It is probably that the cephalon was of equally low convexity, and with a very wide border, but we have not been able to discover these sclerites. Hence, the species is named informally. However, the deeply parabolic pygidium with huge border feebly demarcated from the pleural lobe, the short axis and dorsal effacement are all good specific characters. All except the last is consistent with an assignment to *Licnocephala* as conceived in this work. A large free cheek lacking any defined border (Fig. 26A) is of appropriate type to belong with this species, but there is no specific evidence for this, and it is tentatively recorded in comparison with that of *L. brevicauda*.

Licnocephala n. sp. B

Figure 26M,O

Material. – Pygidia, PMO 208.213, 208.254, 208.292, CAMSM X 50188.58.

Stratigraphic range. – Isolated collections from Nordporten Member, including H collection, probably low in *P. groenlandicus* fauna.

Discussion. – This very flat, smooth pygidium lacking surface sculpture has poorly defined pleural fields adaxially hardly differentiated from the border. The defined truncate axis is almost as wide at its anterior margin as it is long and extends to just more than half pygidial length. Both axial rings and pleural fields are obscurely furrowed, but two pairs of pleural furrows weakly are curved distally on internal mould. Although the general effacement recalls *Uromystrum affine*, the axis is wider and the border is completely different. No other bathyurelline species compares closely, but open nomenclature is advisable until associated sclerites are definitely associated. Another pygidium of somewhat similar type (Fig. 26N) has a narrower axis and different outline and is unlikely to be conspecific.

Licnocephala? n. sp. C

Figure 26R–U

Material. – Pygidia, PMO 208.275, PMO 208.295; weakly associated cranidium, 223.238; free cheek, PMO 221.254.

Stratigraphic range. – Basal Nordporten Member, top 2 m Bassisletta Member, *Chapmanopyge* fauna.

Description. – Pygidium 55 60% as long as wide, and axis about two-thirds pygidial length and one-third width at most. Axis very prominent, well defined except at tip, tapering regularly, with only lateral parts of first two rings visible dorsally. Narrow pleural fields about half width at most of adjacent axis, quite convex next to axis, laterally sloping down to wide flat, to gently concave border, which is slightly narrower behind the axis than it is posterolaterally. Two pairs of straight pleural furrows cross pleural field only, the second fainter, and first interpleural furrow, and hint of second, extend on to border, the former behind a prominent facet. Terrace ridge sculpture is prominent in this species:

dense on the axis and curving forwards medially, sparser on the pleural fields and border and running slightly oblique to the margin.

Discussion. – We have not been able to definitely associate cephalic parts with this stratigraphically early bathyurelline. It lies at the opposite morphological extreme to *Licnocephala* n. sp. A in having a short pygidium and relatively long axis. It more closely approaches certain *Chapmanopyge* species such as *C. carterensis* Loch, 2007, in general proportions, but it lacks the furrow that marks out the edge of the pleural field that is typical of that genus. However, our assignment is appropriately cautious. An incomplete cranidium is considered to be possibly assignable to the same species (Fig. 26Q) having a low glabella and flat, wide border, but paucity of material makes this association tentative, and it could equally belong with *Licnocephala* n. sp. B. A free cheek of *Licnocephala* type is also figured here (Fig. 26P) from a comparable horizon.

Genus *Punka* Fortey, 1979

Type species. – *Bathyurellus nitidus* Billings, 1865, Cow Head, western Newfoundland, designated Fortey, 1979.

Discussion. – There is some dispute as to whether *Punka* can be distinguished from *Licnocephala* Ross, 1951 (Brett & Westrop 1996). The type species of *Licnocephala*, *L. bicornuta* Ross, 1951, is known from rather small silicified examples from the western USA. Nonetheless, Ross (at first tentatively) assigned a characteristic pygidium, which has a wide, flat border, on to which the pleural furrows extend feebly, if at all. Hintze (1953) described a second species, *L.? cavigladius*, of similar age showing some of the same characteristics. In the Spitsbergen section, there are bathyurellids with two clearly separate pygidial morphologies within the Nordporten Member of the Kirtonryggen Formation that we would be reluctant to place in the same genus. The pygidia here attributed to *Licnocephala* are typified by a poorly demarcated, flat pygidial border without clear pleural furrows extending on to it. Pygidia of *Punka flabelliformis* show distinct furrows extending almost to the margin. Despite the general rule that well-developed pleural furrows are likely to be a primitive character, in the case of *Punka*, we believe that the extensive furrows are a derived character. They extend over a wide doublure, which almost reaches the axis. The more plesiomorphic state is shown in several stratigraphically earlier Bathyuridae such as *Peltabellia* (e.g. Ross 1951, pl. 17; Fortey & Peel 1990; *P. glabra* herein) and *Strigigenalis derbyi* Loch, 2007, in which, as in many trilobites, the pleural furrows stop at the inner edge of the pygidial doublure. However, in *P. peltabella,* the pygidial border slopes downwards, rather than being flattened and horizontal, the latter being probably a derived character. If this interpretation is correct, the pygidial differences between *Punka* and *Licnocephala* are likely of real significance. *Chapmanopyge* is another more or less plesiomorphic genus typified by a sharply defined inner margin of the pygidial border. More problematic is the place of *Grinnellaspis* Poulsen, 1948, of which the type species is known only from a pygidium from Ellesmere Island. The name was revived by Boyce (1989) and used again by Brett & Westrop (1996) for trilobites with very straight furrows crossing the pygidial border and radially arranged like the spokes of a bicycle wheel. *Punka akoura* Loch, 2007, shows the same pygidial structure, but with pleural furrows in addition proximally. This genus would then share a derived character of *Punka*. As Brett and Westrop point out, cephalic characters do not 'sort' in an obvious way. Free cheeks are not known for some species attributed to *Grinnellaspis*. Boyce (1989) suggested that *Punka* may have been derived from *Grinnellaspis*, which would imply that together they constitute a clade. However, a new species described below, *Punka latissima*, is clearly related to *P. flabelliformis,* but the furrows crossing the pygidial border are weak compared with that species, and it is plausible that this species illustrates the origin of this character. In this case, *Grinnellaspis* with its powerful pygidial furrows is more likely to show a derived state. Fortey (1979) placed a pygidium of *Grinnellaspis* type into *Punka* (his pl. 35, figs 12, 14, 15). The possibility then remains that this genus will be a senior name for *Punka*, although it would be unwise to follow this without more knowledge of the type species. Further, if Boyce (1989) is correct in assigning cephalic parts to *Grinnellaspis? newfoundlandensis* Boyce, they differ from both species of *Punka* described below in having a long (sag.) convex pre-glabellar field and a free cheek lacking the distinctive furrows separating the genal field from the genal spine. Since pygidial morphologies of *Punka, Licnocephala* and *Chapmanopyge,* seem to be maintained among several species through several successive faunas, it does seem that they are likely to constitute separate clades with separate histories, that is, they merit generic recognition.

Punka latissima n. sp.

Figure 27

Holotype. – Pygidium, PMO 223.237 (Fig. 27J); top bed of Bassisletta Member.

Material. – Cranidia, PMO 208.267, 223.183, 223.217, CAMSM X 50188.36; free cheeks, PMO 208.278, 223.219; pygidia, PMO 208.086, 208.169, 223.204–5, 223.211, 223.235, 223.240, PMO NF 1894, CAMSM X 50188.38.

Stratigraphic range. – Bassisletta Member, top 34 (216–250 m from base), and Nordporten Member to 105 m from base. *Chapmanopyge* to *Petigurus groenlandicus* fauna.

Diagnosis. – *Punka* species with axial sculpture of very dense terrace ridges and sharply defined cephalic border. Relatively wide pygidium with gently sloping, but convex border. Pleural furrows very deep on pleural fields and especially at their margins where inflected backwards, but very weak over border.

Etymology. – Latin: 'most wide'.

Description. – Bathyurelline of medium size with thin cuticle, many specimens partial and broken. Nonetheless, it is a species clearly distinct morphologically and stratigraphically from *P. flabelliformis* and can be named formally. Cranidium convex (sag., tr.), sloping down to sharply delimited border. Glabella 1.5–1.6 times longer than width at midlength, with a bullet-shaped outline in dorsal view, slightly wider at occipital ring, moderately vaulted (tr.) and lacking defined lateral glabellar furrows. Occipital ring well defined by narrow furrow, about one-fifth of total glabellar length. Axial and pre-glabellar furrows narrow and of constant width. Strongly curved palpebral lobes more than one-third (exsag.) length of glabella, horizontally disposed, in front of which pre-ocular fixed cheeks slope downwards steeply and are gently convex to border. Pre-glabellar field very narrow medially, tucked beneath front of glabella, which can be sub-acuminate. Border sharply defined and concave, or flat and angled upwards, about same length (sag.) as occipital ring. Facial sutures diverge moderately strongly in front of palpebral lobes (about 30° to sag. line in dorsal view), more strongly behind eyes at narrow (exsag.), transverse post-ocular cheeks, though preservation imperfect here. Surface sculpture on glabella very

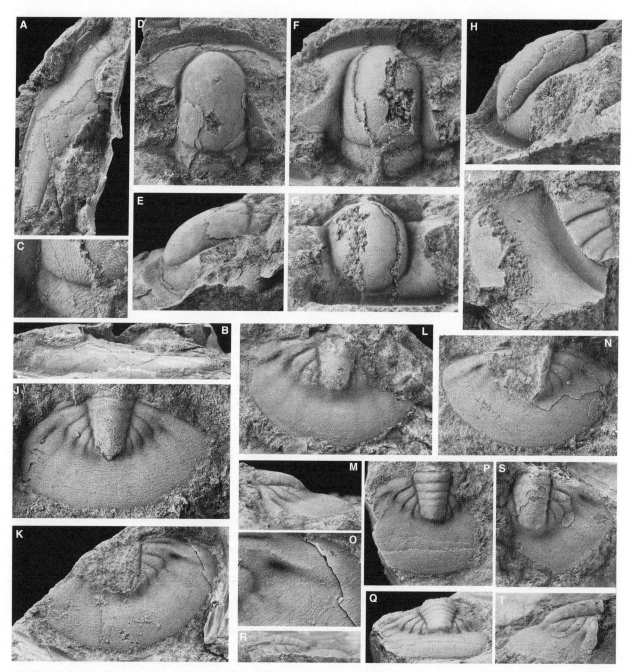

Fig. 27. Punka latissima n. sp. Top 2 m of Bassisletta Member, *Chapmanopyge* Fauna, unless otherwise stated. **A–B**, Free cheek PMO 223.219 in dorsal and lateral views (×3). **D–E**, Cranidium CAMSM X 50188.36 in dorsal and lateral views (×2.5), *P. groenlandicus* fauna, unspecified horizon. **F–H**, Cranidium PMO 223.183 in dorsal, anterior and lateral views (×6), *P. groenlandicus* fauna, 105 m, and (**C**) enlargement of **F** showing dense sculpture. **I**, Pygidial doublure PMO 223.235 in dorsal view (×3). **J**, Latex cast of external mould of holotype pygidium PMO 223.237 in dorsal view, top Bassisletta Member. **K**, Largest, incomplete pygidium PMO 223.211 in dorsal view (×2). **L–M**, Pygidium PMO 208.169 in dorsal and lateral views (×3), 95–108 m, showing weak radial furrows. **N**, Pygidium PMO 208.086 in dorsal view (×4), 106 m. **O**, Enlargement of **J** showing sculpture. **P–R**, Pygidium PMO 223.204 in dorsal, posterior (×4) and lateral (×3) views. **S–T**, Pygidium PMO 223.240 in dorsal and lateral views (×2.5), top Bassisletta Member.

densely disposed lines or ridges, scattered and weak on palpebral lobes and weak or absent on pre-ocular cheek and border, except for fine raised line running around very narrow anterior rim.

Free cheek blade like and four times longer than wide, relatively flat except anteriorly where continuing downward slope of pre-ocular fixed cheek. This adaxial convex area bounded on outer side by shallow furrow that is the true lateral border furrow, which divides off the long, acutely triangular genal spine. Cranidial border continued into concave, relatively wide anterolateral border on cheek, which narrows along the edge of the genal spine, most noticeably opposite posterior end of eye. Sculpture of chevron-shaped terrace lines on distal part of genal spine.

Pygidium up to 1.75 times wider than long, with wide border similar to or slightly less in sagittal length than pygidial axis. Tapering axis with four moderately defined rings slightly decreasing in width (sag.) backwards and abruptly truncate sub-rectangular terminal piece longer than the first ring (sag.). Weakly convex pleural field about two-thirds width (tr.) of axis at it widest, demarcated from the border by a change of slope, but not an incised furrow. First short pleural furrow transverse. Posterior three pairs strongly hooked backwards distally along sloping lateral edge of pleural field. Their continuation onwards over the border is variable. Several specimens show faint, radially disposed furrows continuing across border but not reaching its margin (Fig. 27L). The furrow at the back of the facet is deeper, especially as it approaches the edge of the pleural field. Some specimens are almost smooth bordered. There is no reason to suppose that more than one species is present, and it is considered that these populations show the beginnings of the furrows that become more distinct in *P. flabelliformis*. Border has a distinct, but variable convexity in profile behind the axis. Doublure (Fig. 27I) has strongly upturned inner edge at the base of which there is a broadly depressed band and widely spaced terrace lines on venter sub-parallel to margin. Crowded terrace Bertillon on axis like that on glabella, but sparser, undulating, approximately transverse terrace lines on border.

Discussion. – This species is closely related to *P. flabelliformis*, sharing among other characters the long blade-like genal spine, bullet-shaped glabella with dense terrace sculpture and fan-shaped pygidium carrying inflected pleural furrows, although their continuation on to the border is much weaker in *P. latissima*, and we regard this as the inception of this character in the stratigraphically earliest species of the genus. *Punka latissima* differs unequivocally from *P. flabelliformis* in the density of its axial sculpture, sharper definition of cephalic border, wider pygidium with a border that is convex behind the axis and is approximately similar in length to the axis on the sagittal line. The sharper posterior deflection of the pleural furrows at the edge of the pleural field enables even small fragments to be identified. *Punka latissima* has a long stratigraphic range, but we have not detected differences between the extremes. Another species name that should be considered is *Niobe groenlandica* Poulsen, 1937, based on an incomplete pygidium from the Kap Weber Formation. Clearly, this is a bathyurelline, and the

pleural furrows are similarly disposed to the species described here, so it should be placed in *Punka*. Moreover, it is possible that the free cheek assigned to *Bathyurellus teicherti* by Poulsen (1937) also from the Kap Weber Formation might belong to this species. However, the material is so fragmentary that we are reluctant to place the Spitsbergen material within *Punka groenlandica*, while recognising that more and better material of the latter might establish its priority in the future. However, the pygidial pleural fields of the Greenland specimen do show evidence of strong, ridged sculpture along the pleural lobes, which we have not observed on any of our specimens.

Punka flabelliformis Fortey, 1979

Figure 28

1865 *Bathyurellus marginiatus* Billings, p. 264 (*pars*), fig. 249, *non* fig. 248.

1979 *Punka flabelliformis* Fortey, pp. 96–99, pl. 33, figs 1–10.

1986 *Punka flabelliformis*; Fortey, p. 20, pl. 1, figs 3, 7.

Material. – Cranidia, CAMSM X 50188.72, 50188.87; free cheeks, CAMSM X 50188.49, X 50188.83; pygidia, PMO 208.099, CAMSM X 50188.44, X 50188.82.

Stratigraphic range. – Upper part of Nordporten Member, 150–200 m, *Petigurus nero* fauna.

Discussion. – A full description of this distinctive species from the Catoche Formation of western Newfoundland was given by Fortey (1979) and need not be repeated here, especially since most of its features are similar to those of *P. latissima* above, which it follows in the higher part of the Kirtonryggen Formation. It is, however, illustrated fully to prove its identity. Particularly characteristic of the species is the surface sculpture of strong terrace ridges on the long (sag.) glabella and elsewhere (but more spaced than those of *P. latissima*); the wide and relatively flat pygidium compared with *P. latissima* with the axis clearly shorter than the border behind it, and pleural furrows that continue on to the hardly sloping border present a somewhat sigmoidal profile, except the anterior furrows (behind the facet) that do not continue on to the adaxial part of the pygidial pleural fields. All of these features are well shown on the material from Spitsbergen, and

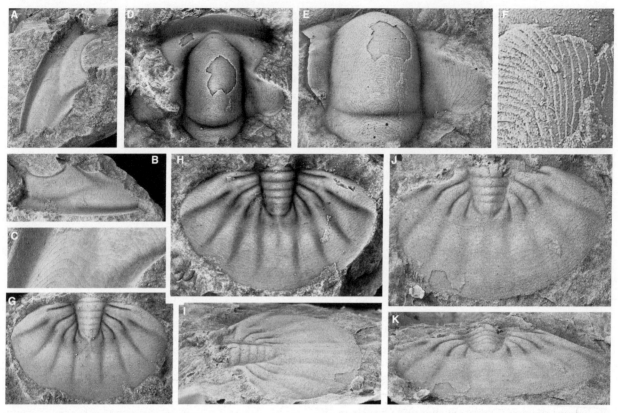

Fig. 28. *Punka flabelliformis* Fortey, 1979. Nordporten Member, *P. nero* fauna. **A–B**, Free cheek CAMSM X 50188.83 in dorsal and lateral views (x2), ca. 150 m, with **C**, enlargement of **A** showing sculpture. **D**, Cranidium CAMSM X 50188.72 in dorsal view (x3), ca. 150 m. **E–F**, Cranidium CAMSM X 50188.87 in dorsal view (x5), unspecified horizon, and **F**, enlargement of **E** showing sculpture. **G**, Pygidium PMO 208.099 in dorsal view (x2) 150 m. **H**, Pygidium CAMSM X 50188.44 in dorsal view (x3) ca. 180 m. **I–K**, Pygidium CAMSM X 50188.82 in lateral, dorsal and posterior views (x4) ca. 150 m.

there are no grounds for separating it from that from the type locality in western Newfoundland. An occurrence of the same species from the Wandel Valley Formation, eastern North Greenland, was illustrated by Fortey (1986), and it may be present in Quebec (Desbiens *et al.* 1996).

Genus *Uromystrum* Whittington, 1953

Type species. – *Bathyurellus validus* Billings, 1865, original designation.

Discussion. – *Uromystrum* is typified by its wide and concave pygidial border, at most weakly crossed by faint pleural furrows. This concave border is underlain by doublure that is closely reflexed against the dorsal cuticle, and it is possible to extract them together forming a semicircle resembling the brim of a hat. Cephalic features differ from the *Punka* complex (above) in having the border furrows effaced so that the genal field extends into a convex and prolonged genal spine without a break. In this feature, *Uromystrum*

resembles *Bathyurellus*, but the pygidium of the latter is otherwise distinctive. The glabella of *Uromystrum* may become effaced in some species, and the cranidial anterior border may become more or less merged with the pre-glabellar field. The convexity (sag.) of the cephalon exceeds that of the pygidium in what is typical illaenimorph morphology.

The situation is complicated by the attribution by Loch (2007, p. 70) of *Bathyurellus affinis* Poulsen, 1937, to *Randaynia*; this is a species previously (Whittington 1953; Fortey 1979; Boyce 1989) placed in *Uromystrum*. Loch points up differences displayed by Poulsen's species as identified by Boyce (1989) from western Newfoundland from the type species of the genus from the Whiterockian boulder at Lower Head. We now have much more material of *affinis* from Spitsbergen, described below. Interpretation of effaced trilobites is notoriously difficult, but overall, the new material does not support the attribution of Poulsen's species to *Randaynia*. The majority of specimens on which the glabellar outline can be seen show it to be medially pointed rather than sub-quadrate, and the tapering proportions of the glabella and its small size in relation to the cranidium as a whole are similar to that of *Bathyurellus* (e.g. *B. diclementsae* n. sp. herein), and unlike *Randaynia*, in which a rectangular glabella occupies much of the cranidium. Effaced though it is, the pygidial axis tapers more rapidly than in *Randaynia* spp., and there is no evidence from published material of the latter of the 'doubled back' doublure we note on *affinis*. However, Loch (2007) is correct in noticing differences from the type species such as its quite well-defined and 'elongate...posteriorly constricted glabella'. We add that the Whiterockian type species does show some furrowing across the concave posterior pygidial border, unlike the species described below. Since furrows crossing on to the border are an advanced bathyurelline character, it is possible that the Middle Ordovician species may have been independently derived, and the concave pygidial border is convergent upon the Ibexian species, as Loch suggested. If we are also correct that *Bathyurellus affinis* Poulsen is not a *Randaynia* species, this opens up the possibility of yet another new, finely drawn bathyurid genus. The presence of a second species closely similar to *affinis* in overlying strata and surely a member of the same clade (*Uromystrum* aff. *U. affine*) might lend support to this view. On the other hand, we also have another distinctive species, *U. drepanon* n. sp., which differs in several other respects from *both* the Whiterockian and Ibexian *Uromystrum sensu lato*. We are reluctant to commit to further sub-division of this group of bathyurellines in this work and here take a broad view of *Uromystrum* pending further data.

Uromystrum affine (Poulsen, 1937)

Figures 29, 30A–I

1937 *Bathyurellus affinis* sp. nov. Poulsen, p. 55, pl. 7, figs 6, 7.

1953 *Uromystrum affinis* (Poulsen); Whittington, p. 660.

1979 *Uromystrum affine* (Poulsen); Fortey, p. 99, pl. 37, fig. 12.

1983 *Uromystrum affine* (Poulsen); Boyce, pp. 173–175, pl. 17, figs 1–4.

1989 *Uromystrum affine* (Poulsen); Boyce, pp. 59–60, pl. 33, figs 7–10, pl. 34, figs 1–4.

2010 *Uromystrum affine* (Poulsen); Boyce and Knight, pl. 3, fig. D.

Material. – Abundant, includes cranidia, PMO 208.223, 208.258, 208.293, 223.119, 223.230, CAMSM X 50188.33a; free cheeks, PMO 208.222, 223.118, 223.223, 223.353 (doublure); pygidia, PMO 221.265, 223.120, 223.132 (doublure), 223.197, 223.210, 223.231, CAMSM X 50188.54, X 50188.88 (+ doublure); partial thorax, CAMSM X 50188.56; hypostomes, PMO 223.153, 223.160, 208.095.

Stratigraphic range. – Nordporten Member of the Kirtonryggen Formation, base to 130 m, *Petigurus groenlandicus* fauna.

Diagnosis. – Modified from Boyce (1989, p. 59). Smooth *Uromystrum* species with weakly defined, narrow, sloping cranidial border, thin marginal rims on triangular free cheeks with short genal spines; pygidium with wide, deeply dish-like border and smooth pleural fields.

Description. – Poulsen's original description of material from East Greenland is brief. Although the species was described further from Newfoundland by Boyce (1989), who associated the several parts, it is represented in Spitsbergen by numerous, well-preserved sclerites, and a fuller description is now possible. Largest cranidia are nearly 3 cm long, so the species was capable of achieving a larger size than *Bathyurellus* spp. Despite this large size, the cuticle is comparatively thin.

Highly convex cephalic shield, curved downwards peripherally, in some specimens making almost a right angle between the occipital edge and the pre-glabellar field. Glabella is faintly defined, though usually recognisable from its gently domed transverse convexity, widest at posterior edge where it exceeds transverse width of adjacent fixed cheek. Gentle forward taper, such that the transverse glabellar width in front of the palpebral lobes is 80–85% maximum width. Because of its deep downward slope, the pre-glabellar field is foreshortened in dorsal view, and the glabella occupies five-sixths of the total length in this orientation (small changes in orientation can alter the proportions). Glabellar length is 1.5–1.6 times maximum width, and axial furrows very shallow, hardly discernible on specimens retaining cuticle. Pre-glabellar furrows converge in a straight course to meet at a low obtuse angle. Glabellar furrows not visible, and occipital furrow just visible on middle part of glabella. Palpebral lobes (exsag.) about 30% cephalic length (sag.), almost semicircular, with no rim and slightly inclined forwards. Anterior end close to axial furrow, with suggestion of an eye ridge; posterior end slightly further away. Posterior section of facial suture transverse

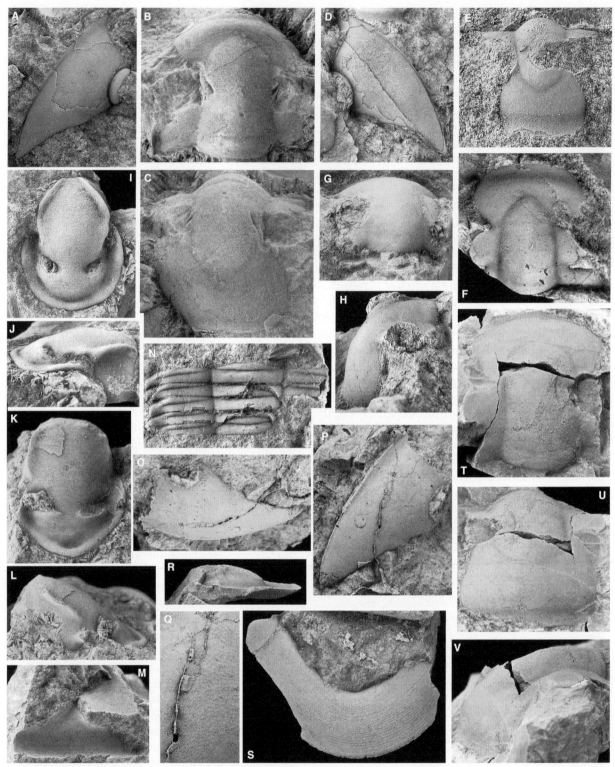

Fig. 29. Uromystrum affine (Poulsen, 1937). Nordporten Member, *P. groenlandicus* fauna. All from 105 m unless stated. **A,** Free cheek PMO 208.222 in near-dorsal view inclined to show eye (x3.5), 130 m. **B–C,** Large exfoliated cranidium PMO 208.293 in dorsal and anterior views (x2), **H** collection (low *P. groenlandicus* fauna, horizon uncertain). **D,** Free cheek PMO 223.118 in dorsal view (x2). **E,** Cranidium with somewhat longer border CAMSM X 50188.33a in anterior view (x7), ca. 105. **F,** Small cranidium with relatively well-defined glabella PMO 223.230, horizon unknown, in dorsal view (x4). **G–H,** Small and highly effaced example of cranidium PMO 223.119 in dorsal and lateral views (x4). **I–J,** Hypostome PMO 223.160 in ventral and lateral views, showing large anterior wing (x6). **K–L,** Hypostome PMO 208.095 in dorsal and lateral views (x4), ca. 105. **M,** Doublure of free cheek PMO 223.253 in lateral view (x2). **N,** Thoracic segments CAMSM X 50188.56 in dorsal view (x3), ca. 105. **O–P,** Free cheek 223.223 in lateral and dorsal views (x1.5), ca. 105, and (**Q**) enlargement of **P** (x6) showing faint caeca. **R–S,** Pygidium broken around doublure CAMSM X 50188.88 in lateral ('Mexican hat') (x1.5) and ventral (x.2.5) views, ca. 105 m. **T–V,** Large, stratigraphically early cranidium PMO 208.258 in dorsal, anterior and lateral views, **H** collection (as **B–C**).

behind palpebral lobe with steep distal curvature to cut margin at acute angle. Post-ocular cheek lacks border furrow, making a strip at least twice as long as wide (exsag.). Anterior section is bowed outwards in a rather uniform arc, then converges adaxially strongly where it crosses the anterior border. The relatively long pre-glabellar field slopes steeply down until it reaches the shallow border furrow, the downward slope continuing more gently on the narrow border to the margin. Dorsal surface lacks sculpture. Internal moulds show evidence of a radial caecal network on the pre-glabellar field, which continues on to the free cheeks.

Free cheek approximates to a triangle, in plan view having a total length twice the width at the eye. Lateral border diminishes in width until it becomes a very thin rim, which, however, continues to the genal angle. Genal field convex and entirely featureless, merging with triangular genal spine which includes an angle of ca. 40° viewed laterally. Eye sits directly on cheek and is about four times as long as high, with impalpably fine lenses. Doublure curves around genal angle, but becomes progressively convex and curved up inside the anterior part of the cheek adaxially. Thus, the anterior part of the cephalon has a doubled back structure, with the inner edge of the doublure lying well above the anterolateral cranidial margin. Doublure carries terrace ridges parallel to margin.

A distinctive hypostome has been associated with this species. It has exceptionally long and wide anterior wings that are directed almost vertically upwards. This makes an appropriate engagement with the cephalic doublure described in the previous paragraph. The middle body is flattened, smooth, but flanked on either side by broad, convex ridges. Deep middle furrows separate a narrow posterior lobe and are directed somewhat inwards and backwards. Posterior border gently rounded about midline.

An incomplete thorax of six segments is almost the only articulated specimen in the Kirtonryggen Formation. It has axial convexity appropriate to this species and low overall convexity (tr.), indicating that the thorax and pygidium together were held well above the level of the cephalic margin. Pleurae are wider than the slightly tapering axis, crossed by quite strong, narrow pleural furrows that do not reach the ends, which are prolonged (slightly backwards) into spinose terminations. Axial rings are about three times as wide as long (sag.) and have prominent half-rings extending up to halfway under the preceding segment. No sculpture seen, other than a raised ridge on the steep, short triangular facet.

Pygidium sub-semicircular or with somewhat more deeply curved posterior outline, maximum width at anterior margin between 55–66% sagittal length. Much of the pygidium is taken up by the distinctly concave border, which is like the brim of a hat, and underlain by closely reflexed doublure. The border + doublure will 'crack out' readily: it is a reinforced cuticle structure, with no space between the upper and lower layers. The inner edge of the doublure curves upwards at the edge of the pleural field and bounded at the top by a slightly swollen rim. Gently convex (tr.) axis is just under half sagittal length of pygidium, tapering to rounded terminal piece. Apart from a defined narrow (sag.) half-ring like those on the thorax, segmentation of the axis is obscure. Similarly, only the anterior segment is defined on the pleural fields, with narrow pleural furrow extending towards proximal part of narrow (exsag.) downturned facet. Up to two further pairs of obscure pleural furrows can be seen on internal moulds on some specimens on the triangular pleural fields.

The species is virtually smooth dorsally. A few terrace ridges are often seen along the posterior edge of the free cheek. The doublure, by contrast, carries on its ventral surface distinct and closely set terrace ridges running parallel with the margin of the exoskeleton.

Discussion. – The type pygidium from East Greenland has been reillustrated by Fortey (1979) and Boyce (1989) and is indistinguishable from the more effaced examples from Spitsbergen. Given other similarities reported here between Poulsen's (1937) Cap Weber fauna and that from the Nordporten Member of the Kirtonryggen Formation, there are no grounds for separating the material described here. The wide, smooth, strongly 'dished' pygidial border is distinctive of the species. *Uromystrum* species typically have deeply convex cephala and weakly convex thoraces and pygidia, with the concave pygidial border variously developed. The type species from the early Middle Ordovician of western Newfoundland differs from *U. affine* in having a well-defined, convex glabella with occipital ring and in having weak pygidial pleural furrows extending on to the border. *U. formosum* (Billings, 1865) from the Cow Head boulders (see Whittington 1963) is most similar to *U. affine* in having a narrow cephalic border, but it has a highly convex glabella and occipital ring. Other Middle Ordovician pygidia seem to be furrowed in similar fashion to that of the type species (e.g. Fortey & Droser 1996) which does open up the possibility that there may be more than one group subsumed in *Uromystrum*. A second species from Spitsbergen, *U. drepanon* n. sp., is also unlike *U. affine* in many features, as mentioned below.

The presence of caeca on the cephalon is matched by *Punka caecata* Fortey, 1980, from the upper part of the Valhallfonna Formation, Spitsbergen. Caeca are often associated with deep-water olenimorph trilobites, but not in these examples. The fact that the present species is associated with a well-oxygenated palaeoenvironment tends to support the view that caeca had nothing to do with living in oxygen-depleted conditions.

Uromystrum aff. *U. affine* (Poulsen, 1937)

Figure 30J–L

Material. – Pygidia, PMO 221.234, 221.264.

Stratigraphic range. – Nordporten Member, 130–190 m from base, top *P. groenlandicus* to *P. nero* fauna.

Discussion. – Rare pygidia generally similar to those of *Uromystrum affine* are found above that species in

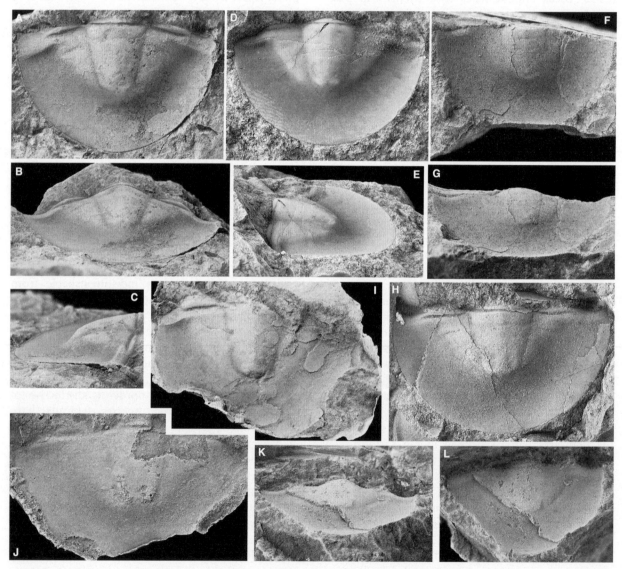

Fig. 30. **A–I**, *Uromystrum affine* (Poulsen, 1937). Nordporten Member, *Petigurus groenlandicus* fauna, 105 m unless noted. **A–C**, Pygidium PMO 223.231 in dorsal, posterior and lateral views (x2.5), **H** collection (low *P. groenlandicus* fauna, exact horizon uncertain). **D–E**, Pygidium CAMSM X 50188.54 in dorsal and lateral views (x2), ca. 105 m. **F–G**, Pygidium PMO 221.265 in dorsal and posterior views (x4), 130 m. **H**, Pygidium PMO 223.120 in dorsal view (x3). **I**, Latex cast of stratigraphically early pygidium showing lateral parts of axial furrows PMO 223.197 in dorsal view (x2), basal *P. groenlandicus* fauna or possibly topmost *Chapmanopyge* fauna. **J–L**, *Uromystrum* aff. *U. affine* (Poulsen, 1937). Nordporten Member, *Petigurus nero* fauna. **J**, Pygidium PMO 221.264 in dorsal view (x4), 130 m. **K–L**, Pygidium PMO 221.234 in lateral and dorsal views (x3), 190 m.

the higher part of the Nordporten Member. Although the definition of the axis on *U. affine* is variable, the pygidial axis is always visible, as it is on Poulsen's type. However, in the specimens under consideration, the axis has merged almost entirely with the adjacent, convex parts of the pygidial pleural fields, presenting a smoothed-out profile. The concave border is also less clearly demarcated from the pleural field. The dorsal surface is entirely without sculpture. We consider that this species is a younger representative of the same clade as *U. affine* and that the differences are sufficient to suggest that it cannot be included as a variation end member of that species. Without recognition of its cephalic parts, we are obliged to record it tentatively as *Uromystrum* aff. *U. affine.*

Uromystrum drepanon n. sp.

Figure 31

Holotype. – Pygidium CAMSM X 50188.43 (Fig. 31P); Nordporten Member, ca. 130 m.

Material. – External mould of cranidium, PMO CAMSM X 50188.21; free cheeks, CAMSM X 50188.45, X 50188.50, X 50188.52; pygidia, PMO 223.191, CAMSM X 50188.47a, X 50188.51, X 50188.73.

Stratigraphic range. – Mid-part of Nordporten Member 105–130 m from base, top *Petigurus groenlandicus* to probable lower *Petigurus nero* fauna.

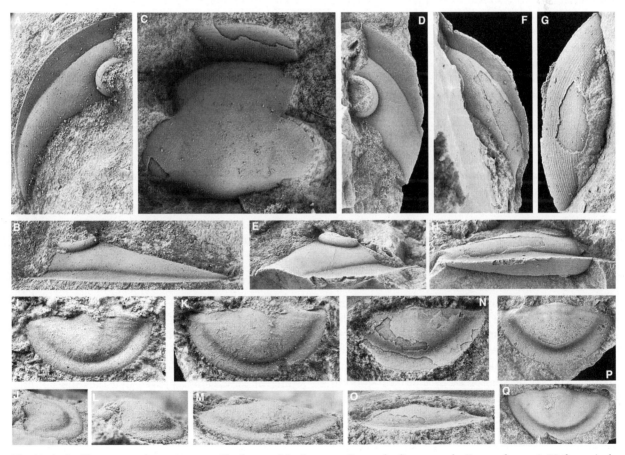

Fig. 31. **A–Q**, *Uromystrum drepanon* n. sp. Nordporten Member, top *P. groenlandicus* to early *P. nero* fauna. **A–M** from single collection, ca. 130 m. **A–B**, Free cheek CAMSM X 50188.52 in dorsal and lateral views (x6). **C**, Latex cast from cranidium CAMSM X 50188.21 in dorsal view (x8). **D–E**, Free cheek CAMSM X 50188.45 in dorsal and lateral views (x4). **F–H**, Free cheek CAMSM X 50188.50 in dorsal, ventral and ventrolateral views showing convex doublure (x4). **I–J**, Pygidium CAMSM X 50188.47a in dorsal and lateral views (x5). **K–M**, Pygidium CAM SM X 50188.73 in dorsal, posterior and lateral views (x5). **N–O**, Pygidium PMO 223.191 in dorsal and posterior views (x5), 105 m, specimen showing flatter and somewhat wider border. **P**, Holotype, pygidium CAMSM X 50188.43 in dorsal view (x3) ca. 130 m. **Q**, Pygidium CAMSM X 50188.51 in dorsal view (x3) ca. 130 m.

Diagnosis. – Generally effaced *Uromystrum* species but with wide, sharply defined cephalic border extending on to remarkably elongate, blade-like genae. Pygidium sub-semicircular, with distinct flat to gently concave, well-defined border. No dorsal surface sculpture, apart from fine terrace ridges on borders.

Etymology. – Greek: *drepanon* – a scythe, referring to the form of the free cheeks.

Description. – This is a distinctive species that does not achieve the large size of other *Uromystrum* species, having pygidia not much more than a centimetre across. It is apparently widespread. Cranidium lacking dorsal furrows, with large palpebral lobe forming an almost perfect semicircle half as long (exsag.) as the border is wide (sag.). Facial sutures diverge at 20–30° in front of eyes, converging in a smooth curve as they cross the border. Posterior section of facial suture shown as outline on free cheek (Fig. 31D), highly divergent and defining narrow spine-like post-ocular cheek. Anterior border abruptly stops the downward slope of the anterior lobe of the cranidium as a horizontal ledge.

Free cheek extraordinarily long and blade like, curving backwards, about 2.5 times the exsag. length at the eye, and readily identified by the border that is defined at its inner edge by the same sharp change in slope seen on the cranidium, with a very narrow border furrow. The genal field is convex, downsloping and carries on without a break to the sharply acute genal angle. Gently dished concave border maintains similar width until about half its length where it begins to narrow progressively towards the genal angle; hence, the proportion of the cheek occupied by the genal field increases posteriorly. The outer edge of the border is marked by a narrow, up-tilted rim. Beneath it, the doublure is convex downwards: hence, the space between genal border and doublure is a flattened tube (see anterior view, Fig. 31H). The doublure carries about 16 prominent raised ridges running parallel to the cephalic margin, which are rather weakly reflected on the internal mould. The dorsal surface is, by contrast, smooth, apart from a few weak terrace ridges running around the edge of the border parallel to the margin, and others oblique to the posterolateral edge close to the genal spine. Eye with minute lenses, strongly curved but not high dorsoventrally, without prominent eye socle, and almost as long (exsag.) as the cheek in front of it.

Pygidium of low convexity, with outline close to semicircular, and prominent border a quarter of the total length on sagittal line. Gently tapering axis feebly defined, effaced posteriorly; anterior of axis recognised by narrow (sag.) half-ring, width there one-third or somewhat less anterior transverse pygidial width. One or possibly two axial rings are faintly discernible. Pleural fields unfurrowed apart from faint half rib. Anteriorly inclined facet narrow (exsag.) and about 20% of anterior width. Border is flattened or very gently convex, fades out before reaching facet, and inner edge is steeply defined by diffuse border furrow, which is deepest medially and closely follows the outline of the posterior pygidial margin, so that the border is of even width along its length. Surface sculpture absent.

Discussion. – The long free cheeks, and the combination of general effacement with sharply defined flat borders, together with the transverse pygidium, make this a particularly distinctive species in the Bathyuridae. It is removed morphologically from *Uromystrum affine*, with regard to the defined borders and distinctive cephalic doublure, and it might be considered to belong to a separate genus. However, the present collections do not have sufficient cranidial material to make this a wise option, and the form of the free cheeks still corresponds with other species of *Uromystrum*, while the wider pygidium compared with other species of the genus is not of great systematic significance. It is considered preferable to extend the compass of *Uromystrum* to include *U. drepanon*.

Pygidia very similar to that of *U. drepanon* are present in the lower part of the Catoche Formation, St George Group, western Newfoundland at Port au Choix (Fortey 1979, pl. 37, figs 10, 11) and from Cape Norman (Boyce 1989, pl. 34, fig. 5).

Family Dimeropygidae Hupé, 1953

Discussion. – Although Whittington (1963) synonymised *Dimeropygiella* Ross, 1951, with *Ischyrotoma* Raymond, 1925, Adrain *et al.* (2001) showed that cladistic analysis of the species previously assigned to either of these genera proved that they did fall into two distinct clades and that both genera could be acknowledged. We follow this interpretation here.

Genus *Ischyrotoma* Raymond, 1925

Type species. – *Ischyrotoma twenhofeli* Raymond, 1925, Lower Head, western Newfoundland, and revised by Whittington (1963).

Discussion. – Adrain *et al.* (2001) retained *Ischyrotoma anataphra* Fortey, 1979, and *I. parallela* Boyce, 1989, both from western Newfoundland, within a restricted concept of *Ischyrotoma* following revision of the group of species to which they are related. However, these species also share a distinctive feature. The cephalic anterior border furrow shallows dramatically over the cranidial section (Fortey 1979, pl. 36, fig. 6). The general rule in trilobites is that deep border furrows continue from free cheek on to the cranidium with a similar depth of impression. To these two species, two more names can be added which show the same feature – *Ischyrotoma sila* Loch, 2007 and *Gelasinocephalus pustulosus* Loch, 2007 (see p. 54). Given its peculiarity, it is likely that this feature is a derived character rather than representing the symplesiomorphic state (cf. Adrain *et al.* 2001). It may eventually prove preferable to separate species showing this feature from both *Ischyrotoma* and *Dimeropygiella*.

Ischyrotoma parallela (Boyce, 1989)

Figure 32A–O

1989 *Ischyrotoma parallela* sp. nov. Boyce, pp. 60–61, pl. 34, figs 7, 8; pl. 35, figs 1–10.

Material. – Cranidia, 223.172, 223.214, 223.233, PMO NF1898, CAMSM X 50188.85; free cheek, PMO 221.239; pygidia, PMO 223.171, 208.245.

Stratigraphic range. – Nordporten Member, 105–120 m from base, *Petigurus groenlandicus* fauna. Also in the H fauna.

Diagnosis. – Modified from Boyce (1989, p. 60). Coarsely tuberculate *Ischyrotoma* with parallel-sided to gently convex-sided glabella with rounded front margin protruding beyond cranidial margin; tuberculate pre-glabellar field present; effaced and inconspicuous cranidial border seen on specimens preserving thick cuticle. Pygidium with prominent, widely separated pair of nodes on terminal piece.

Description. – Although Boyce (1989) described this species from the lower part of the Catoche Formation in western Newfoundland, his material comprised internal moulds, and more details are known from the collections made in Spitsbergen, so a supplementary description is warranted. An internal mould of a cranidium preserved in a manner similar to Boyce's holotype (Fig. 32F) shows deep, nearly parallel axial furrows and a protruding glabella, both

characters in the original diagnosis, and there are not sufficient grounds for establishing a different species herein. However, many Spitsbergen cranidia do show gently convex sides to the glabella, especially on the dorsal surface, so this is regarded as a character variable within the species. Indeed, Boyce's figured specimen (1989, pl. 35, fig. 9) also seems to show this. All cranidia display in profile the front of the glabella protruding to overhang the pre-glabellar field beneath, for example, Figure 32J, and have glabella length/width ratios between 1.45–1.6 (internal moulds measured to centre of wide axial furrows). All also show broadly rounded glabella front in anterior view and below the glabella a distinct pre-glabellar field above the border. Cranidial border itself is sometimes hard to discern, such is the effacement of the border furrow, but is narrow and gently upwardly arched. Most specimens show tubercles continuing across the pre-glabellar field. Boyce's paratype (Boyce 1989, pl. 35, fig. 2) is probably the only specimen in the type series fully to preserve the pre-glabellar area, other cephala breaking off along the doublure. None of the type series preserved the palpebral lobes, which are gently curved, elevated, without tubercles and up to one-third (exsag.) length of glabella. Anterior section of the facial sutures is also somewhat variable, with some specimens initially more divergent than others. Variation in tubercle density is noticeable, but most particularly that the tubercles appear coarser on specimens with cuticle preserved.

A definitely associated pygidium (Fig. 32R–T) is a partial internal mould and not well preserved, but typically dimeropygid with four axial rings and a pair of particularly prominent tubercles at the end of the axis, which are widely separated. Convex bars mark pleural ribs and curve laterally downwards, where their tips are conjoined by a convex, narrow border. A much better-preserved pygidium was collected stratigraphically below the main collection of cranidia figured, but within the range of *Petigurus groenlandicus*, and there is no particular reason to suppose it belongs to a different species. It shows short pleural furrows crossing the proximal parts of the pleural ribs, which carry a few coarse tubercles. The fourth axial ring is completely defined, and the swollen nodes apparently behind that. Rib-like posterior border carries fine raised lines and is upward arched medially.

Discussion. – Adrain *et al.*'s (2001) cladistic analysis paired this species with *I. anataphra* (Fortey, 1979) known from well-preserved material from western Newfoundland and Greenland (Fortey 1986). Both share the unusual feature of an effaced cranidial anterior border furrow, contrasting with a deep lateral border furrow on the free cheek. *I. anataphra* succeeds *I. parallela* stratigraphically in Newfoundland. The latter has a more forwardly jutting glabella that overhangs the anterior border and a rounded, as opposed to acuminate, anterior glabella outline.

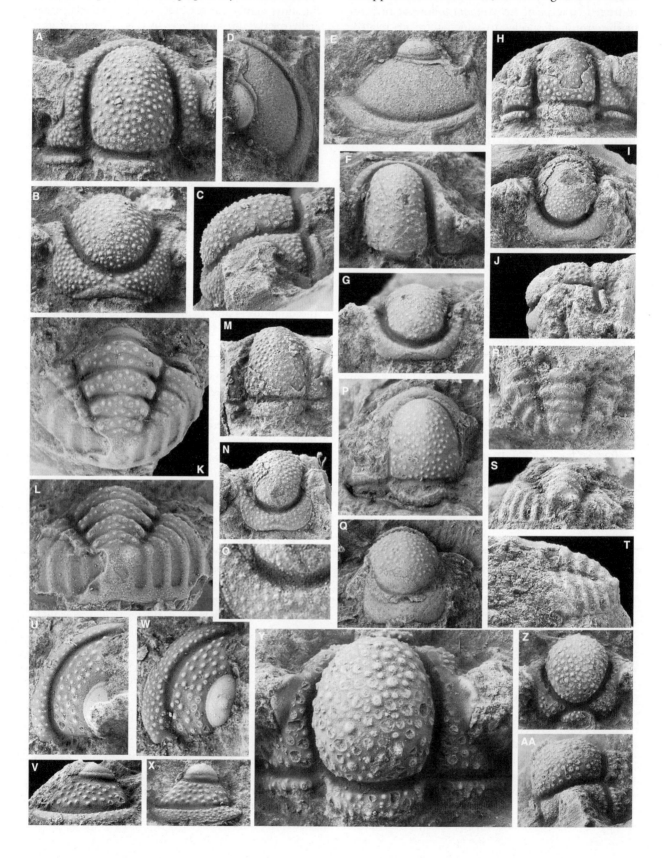

The pre-glabellar field of *I. anataphra* is very narrow medially, and its tuberculation is less coarse. These characters do appear consistent, although the degree of parallel sidedness of the glabella does vary. Fortey's (1979) illustrations of the pygidium of *I. anataphra* show the tubercles on the terminal piece lying side by side, whereas they are widely separated on the pygidium assigned to *I. parallela* herein. Boyce (1989) thought that the two species intergraded stratigraphically, and indeed, the two cranidia figured by Fortey (1979) as 'stratigraphically early' *I. anataphra* show coarser tuberculation and a more anteriorly rounded glabella, but not, apparently, a wider pre-glabellar field. Whether populations really intergrade requires statistical investigation. We have also allowed for a good deal of intraspecific variation in glabellar shape in the interpretation of *I. parallela* here, and it remains possible that there are two closely similar species in the Kirtonryggen Formation. Given the problems with the type material, the conservative approach is to use the prior name *sensu lato*. Loch (2007) described a species from the Kindblade Formation, Oklahoma, as *Ischyrotoma sila*, which is very close to *I. parallela* and would surely belong in *Ischyrotoma* on the criteria of Adrain *et al.* (2001). The cranidial material illustrated of that species is particularly like those specimens of *I. parallela* with more rounded glabella, and the pre-glabellar field is of similar proportions. Perhaps the only difference is in the sculpture, described as 'granular' by Loch, and certainly neither of his illustrated cranidia shows the kind of coarse tubercles we see on the fixed cheeks of *I. parallela*. Given that *I. sila* ranges through three Oklahoma biozones (Loch 2007, pp. 10–11), it would be interesting to know whether some sculptural variation or trend was present through this apparently long interval. Another species from Oklahoma was described by Loch (2007) as *Gelasinocephalus pustulosus*. We have already discussed (pp. 55) reasons why *Gelasinocephalus* seems to bring together a bathyurid (the type species *G. whittingtoni*) with a dimeropygid (*G. pustulosus*). The latter is undoubtedly a species closely similar to our new specimens attributed to *I. parallela* (Boyce 1989), especially those specimens with gently tapering glabellas. *Ischyrotoma pustulosa* (Loch 2007)

cranidia also resemble *I. parallela* specimens from Spitsbergen in having a comparatively long (sag.) pre-glabellar field. The only notable differences are that the fixed cheeks of the Oklahoma specimens seem to lack, or almost lack tuberculate sculpture, which is prominent in our specimens, and in profile, the glabella does not protrude beyond the pre-glabellar field. One Nordporten cranidium (Fig. 32P, Q) is an internal mould that shows a longer pre-glabellar area, no evidence of the anterior border and scattered punctate on the pre-glabellar area. It is recorded as *I.* cf. *parallela* herein and probably represents another species.

Ischyrotoma n. sp. A

Figure 32U–AA

Material. – Cranidium, CAMSM X 50188.13; free cheeks, CAMSM X 50188.14, X 50188.16.

Stratigraphic range. – Top 30 m of Nordporten Member, *Petigurus nero* to *Lachnostoma platypyga* fauna.

Description. – Convex cranidium with anterior border tucked under frontal glabellar lobe. Glabella more or less ovoid in dorsal aspect in front of occipital ring with maximum width at midlength 70% sag. length, and rounded-convex transversely, inflated above occipital ring. All furrows – axial, pre-glabellar, occipital and anterior/posterior border furrows – of similar deep expression. Lateral glabellar furrows lacking. Well-defined occipital ring about 15% total glabellar length. Posterior fixed cheeks triangular, about half (tr.) maximum width of glabella and defined by rather weakly divergent posterior sections of facial sutures (30° to sag. line). Convex anterior fixed cheeks slope steeply downwards and are bounded by outwardly bowed sutures that converge rapidly anteriorly. Palpebral lobes at high point of cheek, about quarter length (exsag.) of glabella, and convex rims gently curved outwards. Convex anterior border not widening greatly

---

Fig. 32. **A–O,** *Ischyrotoma parallela* Boyce, 1989. *P. groenlandicus* fauna. **A–C,** Cranidium CAMSM X 50188.85 in dorsal, anterior and lateral views (x10) exact horizon unknown. **D–E,** Free cheek PMO 221.239 in dorsal and lateral views (x6), H collection, low *P. groenlandicus* fauna, exact horizon uncertain. **F–G,** Cranidium PMO 223.233 in dorsal and anterior views (x5), 105 m. **H–J,** Cranidium PMO 223.172 in dorsal, anterior and lateral views (x5), 105 m. **K–L,** Pygidium PMO 208.245 in dorsal and posterior views (x10), basal Nordporten Member. **M–N,** Cranidium PMO NF 1898 in dorsal and anterior views (x6), 105 m. **O,** Enlargement of **N** showing structure. **R–T,** Pygidium PMO 223.171 in dorsal, posterior and lateral views (x8), 105 m. **P–Q,** *Ischyrotoma* cf. *I. parallela* Boyce, 1989. Cranidium PMO 208.144 (2) in dorsal and anterior views, 105 m. **U–AA,** *Ischyrotoma* n. sp. A. Nordporten Member, all figured specimens from top 3 m *Lachnostoma platypyga* fauna. **U–V,** Free cheek CAMSM X 50188.14 in dorsal (x9) and lateral (x7) views. **W–X,** Free cheek CAMSM X 50188.16 in dorsal (x12) and lateral (x8) views. **Y–AA,** Cranidium CAMSM X 50188.13 in dorsal (x14), anterior and lateral (7) views.

medially, very weakly arched if at all, and in anterior view about the same length as the preglabellar area above it, which is formed by the convergence of deep border furrow and preglabellar furrow: there is no true pre-glabellar field. Coarse tuberculation is typical, also on border and on palpebral rim of one specimen, and among the tubercles a few are noticeably prominent.

Free cheek of usual form for genus, lacking genal spine, and with border narrowing anteriorly (dorsal view). Tubercles on sloping, bevelled border are smaller than those on the genal field. A particular feature of the eye is the prominent tumid eye socle beneath the sub-lunate visual surface, under which the cheek is smooth around the elevated ocular area.

Discussion. – Adrain *et al.* (2001) drew attention to the structure of the anterior border as a way of discriminating *Ischyrotoma* from *Dimeropygiella*: in the latter the cranidial border widens medially seen from above, and in anterior view has a w-shaped profile, compared with an altogether simpler structure in *Ischryotoma*. The species described here seems to belong to *Ischyrotoma*, in so far as manual preparation allows us to see the border. It differs from all other species of early Floian age, but since we lack a pygidium, it is advisable to use open nomenclature. *Ischyrotoma wahwahensis* Adrain *et al.* 2001 has a similar anterior profile, but has an elongate (sag.) glabella compared with *I.* n. sp. A, together with a relatively short eye, and lacks the independently convex eye socle of the Kirtonryggen species. To the list of *Ischyrotoma* species in Adrain *et al.* (2001) must be added *I. sila* Loch 2007 from the Kindblade Formation, Oklahoma, a species discussed above in relation to *I. parallela* Boyce, which is one of a small group of species with an effaced cranidial border. Loch (2007, pl. 28) also figured a number of cranidia of species attributed to *Ischyrotoma* under open nomenclature, but all of them possess a pre-glabellar field and cannot be conspecific with the species from the Nordporten Member.

Family Proetidae Salter, 1864

Genus *Phaseolops* Whittington, 1963

Type species. – *Phaseolops sepositus* Whittington, 1963, Lower Head boulder, western Newfoundland (early Whiterockian).

Discussion. – When Whittington (1963) described the type species of *Phaseolops* from the Whiterockian Lower Head boulder at Cow Head, western Newfoundland, it was the oldest named proetid. Adrain & Fortey (1997) described a still older, late Floian, species from the Tourmakeady Limestone, western Ireland. Both these species occur in carbonate mound, illaenid–cheirurid biofacies. A species from the Nordporten Member of the Kirtonryggen Formation is still older, from the base of the Floian, or even top Tremadocian, and occurs in shallow-water, carbonate platform rocks. Adrain & Fortey (1997) discussed the generic concept of *Phaseolops*. It is difficult to assign these early proetids with confidence, because they seem to display a mosaic of characters, some of them plesiomorphic. In several ways, the new species described below differs from *Phaseolops*, notably in having a glabella that expands in width medially, rather than tapering, lacking a pre-glabellar field, and having a more typically proetid free cheek. While we considered creating a new genus for its reception, we observed the principle that it would not be wise to do so in the absence of a pygidium, which has so far eluded discovery. To reflect this uncertainty, we include the new species with considerable caution in *Phaseolops*.

Phaseolops? bobowensi n. sp.

Figure 34

Holotype. – Cranidium PMO 223.252 (Fig. 34F–H); Nordporten Member, probably 105 m.

Material. – Cranidia, PMO 208.123, 223.165; CAMSM X50188.71; free cheeks, PMO 208.094, 208.156.

Stratigraphic range. – Nordporten Member, base to 105 m from base, *Petigurus groenlandicus* fauna.

Etymology. – For Dr R. M. Owens, distinguished long term student of proetid trilobites.

Diagnosis. – Proetid assigned tentatively to *Phaseolops*, with glabella that is widest medially, concave cephalic anterior border. Posterior border furrow on free cheek fades rapidly laterally. Distinctive surface sculpture of fine ridges, anastomosing and forming maze-like pattern.

Description. – Small species. Characteristic sculpture confidently links cranidium and free cheek and

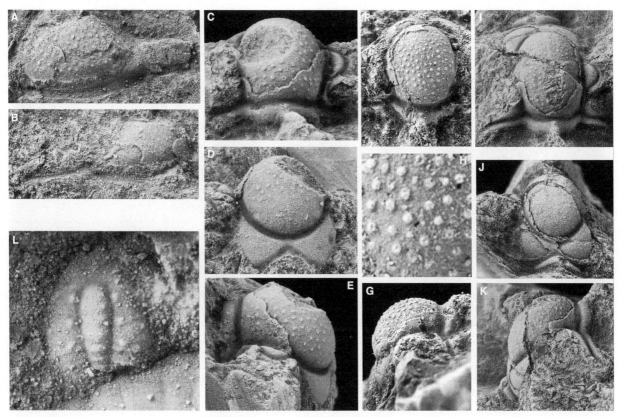

Fig. 33. **A–K**, *Psalikilopsis* n. sp. aff. *P. cuspicaudata*, Ross, 1953. Nordporten Member, *P. groenlandicus* fauna (see p. 52). **A**, Free cheek PMO 221.262 in lateral view (x10), 130 m. **B**, Free cheek PMO 223.227 in lateral view (x10), 105 m. **C–E**, Cranidium PMO 223.184 in dorsal, anterior and lateral views, (x8), 105 m. **F–G**, Cranidium PMO 223.164 in dorsal and lateral views (x10), 105 m. **H**, Enlargement of **F** showing structure. **I–K**, Holotype, cranidium PMO 223.185 in dorsal, anterior and lateral views (x6), 105 m. **L**, Proetoid metaprotaspis PMO 223.242 in dorsal view (x50), uppermost Nordporten Member, *L. platpyga* fauna.

would be expected on pygidial axis; no such pygidium was found. Transverse convexity of glabella is rather uniform along its length, with rounded frontal lobe almost overhanging pre-glabellar field. Maximum width of glabella 60% sagittal length, and occipital ring is about one-sixth of total glabellar length. Axial furrows take a sinuous course and are particularly well defined posteriorly. Pre-glabellar furrow shallowest anteromedially but always visible. Glabella expands in width to a point anterior to the mid-length of the palpebral lobes, which is at just more than half sagittal glabellar length (dorsal view) measured from the back, in front of which the glabella is constricted in width at the point at which the palpebral lobes join the axial furrows. The frontal lobe of the glabella is again wider, making an approximate semicircle in outline. There is no definition of lateral glabellar furrows. A poorly defined smoother area adjacent to the axial furrows before half palpebral length may be a muscle insertion area. Well-defined occipital ring with median tubercle, slightly broader (sag.) medially, and as wide (tr.) as widest part of glabella in front. Narrow (exsag., tr.)

post-ocular fixed cheeks half transverse width of ring, comprised mostly of posterior border. Palpebral lobes posteriorly placed, more than 40% length (exsag.) of glabella, weakly arcuate outline, lacking rims, except anteriorly, where faint palpebral furrow on holotype. Front and back of palpebral lobe almost touch axial furrow. Anterior branch of facial suture weakly divergent at about 20° to sag. line before curving to become almost parallel, outlining sloping, but gently convex, pre-ocular fixed cheeks. Narrow, steeply sloping pre-glabellar field narrowest medially, stopping abruptly at well-defined border, which is half or slightly more width (sag.) of occipital ring in dorsal view. Border rather flat or slightly concave with downturned rim, and even width along its length.

Free cheek prolonged into long, blade-like, triangular genal spine without break, but posterior border furrow fades rapidly at base of spine. Lateral border with upturned rim gets progressively narrow laterally, but rim persists into genal spine, terminating its otherwise steep slope. Posterior border furrow defines flattish, slightly depressed posterior border

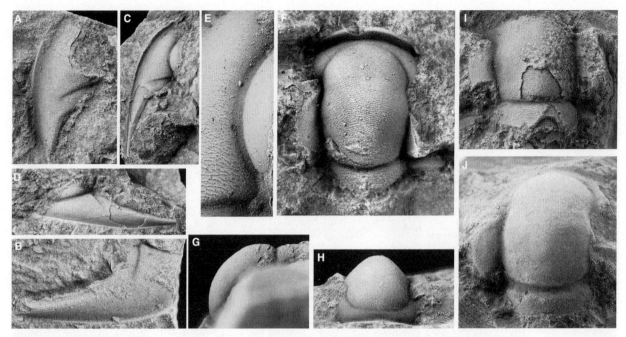

Fig. 34. Phaseolops? bobowensi n. sp. Nordporten Member, *P. groenlandicus* fauna. **A–B**, Free cheeks PMO 208.156 (**A**) dorsal view, 208.156 (**B**) lateral view (x5), basal Nordporten Member. **C–E**, Free cheek PMO 208.094 in dorsal and lateral views (x3), 106 m, with (**E**) enlargement (x10) of **C** showing base of eye and sculpture. **F–H**, Holotype cranidium PMO 223.252 in dorsal (x8), lateral and anterior views (x6), probably 105 m. **I**, Incomplete cranidium CAMSM X 50188.71 in dorsal view (x8), 105 m. **J**, Internal mould of cranidium PMO 208.123 in dorsal view (x6), 105 m.

that merges laterally with base of genal spine. No eye socle, long, curved eye lobe resting on cheek directly.

Sculpture of fine ridges that are not reflected on internal moulds. These ridges are not terrace ridges. They are convex and meandering, sometimes anastomosing, forming a fine-scale maze-like pattern. They are present on the glabella and circum-ocular part of the free cheek, fading out on the pre-glabellar field and border.

Discussion. – It is noted under the generic discussion that the attribution to *Phaseolops* is somewhat arbitrary. The species from the Kirtonryggen differs from the type species from younger Ordovician strata in Newfoundland (Whittington 1963) in its distinctive glabellar shape, lack of incised glabellar furrows and sculpture. *P. ceryx* Adrain & Fortey, 1997, from the Tourmakeady Limestone (western Ireland), has a rather evenly tapering glabella, preglabellar field and very different border and free cheek. It is unfortunate that we were unable to recover a pygidium properly to characterise the species. However, the cranidium and free cheek are distinctive enough to justify the erection of a named new species, especially since this new discovery is the earliest representative of its superfamily.

Family Telephinidae Marek, 1952

Genus *Carolinites* Kobayashi, 1940

Type species. – *Carolinites bulbosus* Kobayashi, 1940, Caroline Creek, Tasmania.

Carolinites? n. sp. A

Figure 35

Material. – Cranidia, PMO 208.085, 223.203, 223.212, 223.213, 223.246; free cheek fragment, PMO 223.229.

Stratigraphic range. – Recovered from the topmost 2 m of the Bassisletta Member (topmost *Chapmanopyge* fauna) and the lower Nordporten Member to 105 m from base, *Petigurus groenlandicus* fauna.

Diagnosis. – A *Carolinites* species with characteristically shaped, obtusely rounded glabella and narrow cranidial border, but lacking bacculae on posterior

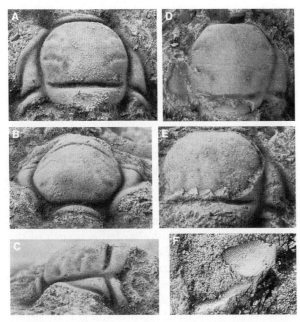

Fig. 35. *Carolinites*? n. sp. A. Topmost 2 m Bassisletta Member, *Chapmanopyge* fauna and Nordporten Member 105 m from base, *P. groenlandicus* fauna. **A–C,** Cranidium PMO 223.203 in dorsal, anterior and lateral views (x10), top 2 m. Bassisletta. **D,** Cranidium PMO 208.085 in dorsal view (x7), Nordporten Member, 106 m. **E,** Cranidium PMO 223.213 in dorsal view (x10), Nordporten Member, 105 m. **F,** Free cheek fragment PMO 223.229 in dorsal view (x14), top 2 m Bassisletta Member.

adaxial corners of fixed cheeks, and with two pairs of glabellar furrows.

Description. – This interesting species is the stratigraphically earliest of the genus, but is known imperfectly, with no pygidium assigned, so although it is unquestionably new, it has to be described under open nomenclature. There are subtle differences between the stratigraphically earliest collection, which includes well-preserved cranidia, and that from 105 m above the base of the Nordporten Member, where preservation is not as good. The best-preserved cranidium has glabella as wide as long, expanding in width to a point shortly in front of its mid-length, whence it narrows into very broadly rounded anterolateral corners before being transverse across mid-line. The transverse convexity of the glabella is squatter than is typical of most species of the genus, a feature particularly evident in the larger cranidium from the same horizon as the best-preserved cranidium (Fig. 35D). A deeply defined occipital furrow is transverse, fading at extreme lateral edges, defining band-like occipital ring, comprising one-quarter glabellar length (dorsal view). Two pairs of glabellar furrows are not deeply incised, but clear depressions, the first pair at mid-length of pre-occipital glabella, arched evenly inwards and

backwards not quite to a third glabellar width, the anterior pair fainter but parallel. Neither quite reach the deep, narrow axial furrows. Narrow (tr.) fixed cheeks are at their widest less than one-third max. glabellar width, with very long palpebral lobes running their whole length. There are no bacculae. Palpebral lobes are well defined, inflected behind their mid-length, down-turned anteriorly. Posterior cranidial border well defined by deep furrow, sloping outwards and backwards. Anterior cranidial border convex, narrow (sag.), arched upwards medially where it is widest. No surface sculpture seen. A small free cheek has the expected large eye, little in the way of genal field and apparently long genal spine. Specimens from the higher horizon (Fig. 35E) are not so well preserved, but have a more wedge-shaped glabella, with less rounded anterolateral corners of the glabella, and the fixed cheeks may be proportionately narrower.

Discussion. – McCormick & Fortey (2002) noted that the bacculae typically developed at the inner posterior corners of the fixed cheeks of *Carolinites* were least developed in the earliest species, *C. nevadensis* Hintze, 1953, from Utah, Spitsbergen and western Newfoundland, and became larger in younger species. The species described above is still older and lacks bacculae altogether. The stratigraphically younger specimens of *C.* ? n. sp. A have the sub-quadrate, wedge-like glabella typical of the genus, but the earlier specimens are more rounded at the glabellar anterolateral corners. It may be that more than one species – an earlier and a later one – are represented in our collections, though the material is inadequate to prove it. Although there are similarities to younger *Carolinites* spp., the presence of distinct glabellar furrows is an important difference. The much younger *Carolinites rugosus* Fortey, 1975, from the Profilbekken Member of the Valhallfonna Formation displays complex glabellar muscle impressions, but they are not particularly similar to those of *C.*? n. sp. A, and that species also has a complex surface sculpture.

We regard it as more likely that this species is related to an unusual trilobite from the Olenidsletta Member of the Valhallfonna Formation described by Fortey (1980a) as *Buttsia inexpectata*. The attribution of this late Ibexian species to a Cambrian genus and family (Catillicephalidae) was rather implausible, although nothing like it had been described from contemporary strata. It is clearly like *Carolinites*? n. sp. A in having a wide, wedge-shaped glabella with broadly backward curved glabellar furrows, although its palpebral lobes are shorter than the species discussed here, and the glabella outline differs. It is certainly the case that the new, earlier species from the Kirtonryggen Formation is a typical telephinid. It seems quite likely that there is another rare Early Ordovician genus allied to *Carolinites*, but with defined glabellar furrows, to which both *buttsi* and n. sp. A might be assigned. In neither case do we fully know the cephalic and pygidial sclerites, and we are obliged to use open nomenclature at this stage of knowledge.

Order Corynexochida Kobayashi, 1935
Family Illaenidae Hawle & Corda, 1847

Genus *Illaenus* Dalman, 1827

Subgenus *Illaenus* (*Parillaenus*) Jaanusson, 1954

Type species. – *Illaenus fallax* Holm, 1882, p. 82, Chasmops (probably Dalby Limestone) and Kullsberg limestones, Sweden, original designation.

Illaenus (*Parillaenus*) *primoticus* n. sp.

Figure 36

Holotype. – Cranidium, PMO 208.073 (Fig. 36E–F); Nordporten Member, 106 m.

Material. – Cranidia, PMO 208.075, 221.283, 221.285, NF.1897; free cheeks, PMO 208.082, NF 1896; part thorax of seven segments with attached pygidium, PMO 208.070; pygidia, PMO 208.072, 208.074, 208.080, 208.087, 221.284, 221.286, 221.287, 221.288, 223.133.

Stratigraphic range. – Lower Part of the Nordporten Member, 95–106 m from base, *Petigurus groenlandicus* fauna.

Diagnosis. – Exoskeleton with weakly defined terrace lines on all parts. Cranidium with faint dorsal furrows, lunette lacking. Palpebral lobe approximately midway between anterior and posterior margins, free cheek elongate in dorsal view with broadly rounded lateral margin. Pygidium with ill-defined rachis, inner edge of doublure entire, faint median notch with posterior furrow.

Etymology. – Latin: for dawn – appearing early.

Description. – Isolated exuviae of the species occur in large numbers at the 106-m level (referred to as 'the illaenid bed' in the field), but are less common in beds immediately below and above. Specimens fall in the small range with cranidia and pygidia both between 3–15 mm in transverse width. The exoskeleton is thin and easily exfoliates.

Cranidium, moderately convex, glabella best defined on the smallest specimens and effaced on the larger except for the posterior end where it is defined by the notched termination of the dorsal furrows in the posterior margin. Where exfoliated, the furrow is not visible on the internal mould, and there is no lunette. A median glabellar tubercle is also lacking.

The posterior margin is curved outwards and backwards from the end of the dorsal furrow and is pointed where it meets the posterior branch of the facial suture. When the cranidium is seen in palpebral view, the lobe is situated at mid-length between the anterior and posterior margins and is not separated from the fixed cheek by a change in slope. Posterior branch of facial suture straight to gently curved backwards, anterior suture at first straight and curved gradually inwards to the anterior border outside a projected line through the dorsal furrow. Genal angle of free cheek broadly rounded oblique notch in lateral border in line with middle of eye lobe. Faint terrace lines tail off from maximum number on cheek surface anterior to eye towards genal angle and here surface smooth. Terrace lines continue on frontal part of cranidium, but are absent dorsally on glabella and fixed cheeks behind palpebral lobes.

Details of rostral plate and hypostoma not known.

Pygidium boat shaped, rachis faint to effaced. Paradoublural line indicated by change in slope which is vertical posteriorly and steeply sloping towards long, acute, articulating facet. Doublure, broad with smooth inner margin save for faint median depression and furrow extending approximately halfway down the posterior slope of the doublure. Terrace lines parallel to margin laterally, becoming gently curved upwards at mid-line and strongly curved into median furrow.

Thorax segment number not known, but the only specimen known with partial thorax shows parts of seven segments attached to a pygidium. The thorax tapers slightly backwards and is moderately convex with a maximum transverse width at the anteriormost preserved segment slightly more than twice the width of the adjacent pleura. Latter horizontal, curved gently downwards distally with narrow facet. Faint terrace lines behind facet, otherwise pleurae and rachis smooth.

Discussion. – To our knowledge, *I.* (*Parillaenus*) *primoticus* n. sp. is the earliest described illaenid.

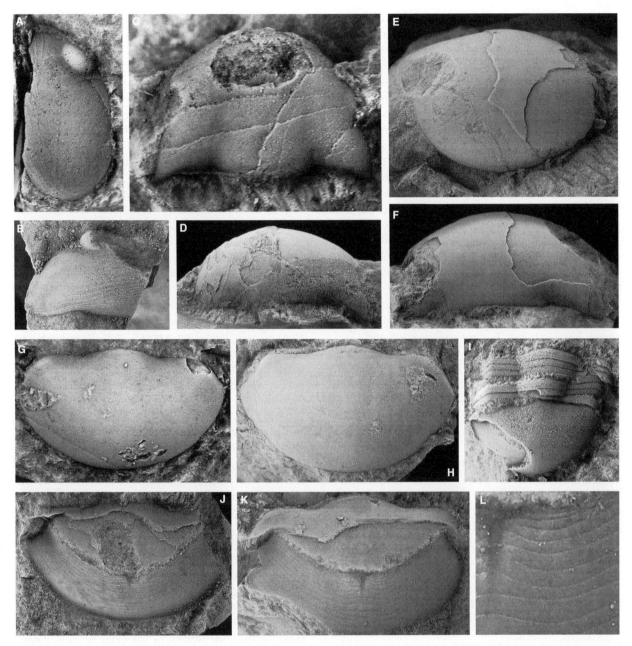

Fig. 36. Illaenus (Parillaenus) primoticus n. sp. Nordporten Member, *Petigurus groenlandicus* fauna, all 106 m. **A**, Free cheek PMO NF 1896 in plan view (x7). **B**, Free cheek PMO 208.082 in anterior view (x5) showing terrace lines. **C**, Juvenile cranidium PMO NF 1897 in dorsal view (x14) showing palpebral lobe on right. **D**, Large cranidium PMO 208.075 in dorsal view (x3). **E–F**, Holotype, cranidium PMO 208.073 in anterior and dorsal views (x4). **G–H**, Pygidium PMO 223.133 in dorsal view (x5). Pygidium PMO 208.074 in dorsal view (x5). **I**, Thoracic segments and pygidium, PMO 208.070 in dorsal view (x5). **J**, Pygidium with doublure prepared, PMO 208.087 in dorsal view (x5). **K–L**, Pygidium PMO 208.080 in posterior view to show doublure (x5) and (**L**) enlargement (x15) showing terrace lines.

Boyce (1989, p. 33) mentioned but did not describe an *Illaenus* sp. from the lower limestone member of the Catoche Formation, western Newfoundland. In view of the other similarities between Spitsbergen and the St George Group faunas, this species may well prove conspecific. The genus *Illaenus* reappears in the Spitsbergen succession in the Profilbekken Member of the Valhallfonna Formation, where *I. oscitatus* Fortey, 1980, is distinguished among other features by its characteristic punctate sculpture.

Following discussions by Owen & Bruton (1980) and Bruton & Owen (1988), the species is assigned to the sub-genus *Parillaenus* Jaanusson, 1954, by reason of the simple, curved inner margin of the pygidial doublure as opposed to the cuspate doublure so common in *Illaenus* s.s. and *Stenopareia*. The significance of this feature was discussed at length by Begg (1945) and Jaanusson (1954), although Whittington (1963, p. 68) did not regard it as an adequate feature for separating groups of *Illaenus* species. Species of *I. (Parillaenus)* are known from the Sandbian to Hirnantian in Scandinavia, whereas those with a cuspate pygidial doublure, centred around *Illaenus sarsi* (see Jaanusson 1957, pls. 4–6), are Darriwilian and younger. In North America are species of *Illaenus* from rocks of Darriwilian age in Western Newfoundland (Whittington 1963, 1965), California and Nevada (Ross (1967, 1970), all of which share features with *Illaenus s.s.* rather than *Parillaenus*. These include the cuspate inner margin of the pygidial doublure and axe-like rostral flange as illustrated for *I. tumidifrons* Billings, 1865 (see Whittington 1963, pl. 15, fig. 11 and pl. 16, fig. 16), and *I. marginalis* Raymond, 1925 (see Whittington 1965, pl. 47, fig. 10), *Illaenus auriculatus* Ross (1967, pl. 5; 1972, pl. 12; pl. 14, figs 23–25) and general features such as the lunette and glabellar tubercle on *I fraternus* Billings, 1865 (see Whittington 1965, pl. 45, figs 1, 4), *I. cf. I. alveatus* Raymond and *Illaenus* sp. 1 (Ross 1970, pl. 15). Similar features of the pygidial doublure and rostral plate are present in *I. weaveri* Reed from the Tourmakeady Limestone (late Floian) of western Ireland (Adrain & Fortey 1997, pl. 5; pl. 6, 1–12). However, *I. weaveri*, like *I. (Parillaenus) primoticus* n. sp., has a markedly effaced cephalon, and the lunette is lacking on both dorsal and ventral surfaces. This latter feature is otherwise characteristic of our early illaenid.

Family Leiostegiidae Bradley, 1925

Genus *Leiostegium* Raymond, 1913

Type species. – *Bathyurus quadratus* Billings, 1860.

Remarks. – The question of the distinction (or not) between *Lloydia* Vogdes, 1890, and *Leiostegium* has been discussed by several authors (e.g. Ross 1970; Boyce 1989), but cannot really be resolved without redescription of the type species of both of them. *Leiostegium* has generally been used for Early Ordovician leiostegiid species, and this practice is followed in the present work.

Leiostegium spongiosum n. sp.

Figure 37

Holotype. – Pygidium, PMO 208.058 (Fig. 37G–I); Spora Member.

Material. – Among additional material: cranidia, PMO 208.178, CAMSM X50118.61; free cheek, CAMSM 50118.63; pygidia 208.179, 208.190, CAMSM X50118.68.

Stratigraphic range. – Spora Member (undivided) of the Kirtonryggen Formation, early Ibexian, Stairsian, *Svalbardicurus delicatus* fauna.

Diagnosis. – *Leiostegium* with fine sponge-like surface sculpture over most of the dorsal surface. Glabella outlined by axial furrows that are only slightly concave medially Cranidial border very narrow in front of glabella. Pygidial axis with seven rings, of which the first five are well defined.

Description. – Cranidium moderately convex (sag., tr.), glabella standing above cheeks, and with a broadly rounded transverse profile. Glabella with maximum transverse width at occipital ring twice that of adjacent fixed cheeks, and two-thirds, or slightly more glabellar length in dorsal view. Glabellar sides are concave, but rather weakly so, narrowest glabella opposite the anterior edges of the palebral lobes. Glabellar furrows indicated by smooth patches interrupting sculpture, of which the posterior pair are most clearly shown, being approximately triangular and widening inwards; small, sub-circular impressions close to the axial furrows are located behind the point at which the eye ridges join the axial furrows. Occipital furrow deep, defining occipital ring which is widest medially and has indications of a pair of muscle impressions laterally on its anterior edge. Glabella protrudes on to border, where it is rounded gently and evenly about the mid-line. Pre-glabellar furrow has about the same depth as the axial furrows. However, the axial furrow becomes shallower where it abuts the border furrow, and where it shallows, a small tubercle is present in the axial furrow. Palpebral lobes are medially placed, rims somewhat elevated, 25% or somewhat less (exsag.) glabellar length. Eye ridges clearly visible but scarcely prominent, running inwards and forwards. Pre-ocular cheek defined by very gently divergent anterior branch of facial suture, hence widening a little anteriorly. Post-ocular cheek triangular, defined anteriorly by posterior branch of facial suture at 40° to sagittal line. Posterior border furrow deep and widest abaxially; posterior border narrow and convex. Anterior border convex and forming a ledge, half as wide (sag.) where glabella encroaches upon it. Surface of cuticle is mostly covered by a very fine pitted sculpture giving surface a spongy appearance, with tiny granules additionally in genal

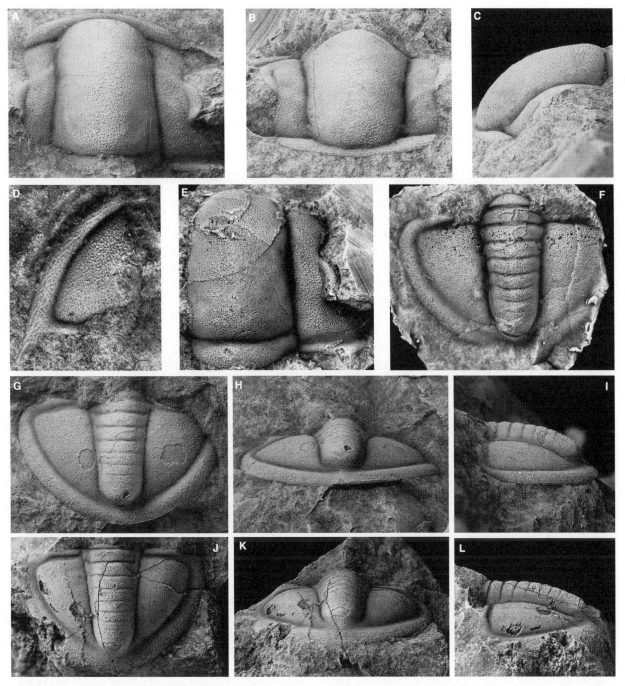

Fig. 37. *Leiostegium spongiosum* n. sp. *S. delicatus* fauna, all Spora Member (1–20 m, undivided horizons). **A–C**, Cranidium PMO 208.178 in dorsal, anterior and lateral views (x9). **D**, Free cheek CAMSM X 50188.63 in dorsal view (x10). **E**, Cranidium CAMSM X 50188.61 in dorsal view (x5). **F**, Latex cast of pygidium CAMSM X 50188.68 in dorsal view (x5) with stronger axial ring furrows. **G–I**, Holotype pygidium PMO 208.058 in dorsal, posterior and lateral views (x5). **J–L**, Pygidium PMO 208.190 in dorsal, posterior and lateral views (x3).

region. Border has raised lines along its anterior edge.

Free cheek has an outline of a quarter circle, its genal field also covered with fine pitting which extends on to the inner edge of lateral border, which carries oblique ridges on its outer part. Both lateral and posterior border furrows are deep and hardly shal-

low towards the genal angle, meeting at somewhat less than a right angle. Eye surface not seen. There is a short pointed genal spine.

Pygidium 0.65–0.75 as long as wide, the axis at its widest about 30% maximum pygidial width just behind anterior lateral corners. Axis is slightly funnel shaped, reaching border furrow, and somewhat variable in length/width, but with seven sub-equal axial rings of which the first five usually are narrowly defined almost to the axial furrows, the last two less so, and fainter. The rounded terminal piece is quite convex, equal in length to about two rings together. The convex border is of the usual, even width type for the genus. Border furrow weaker behind axis. Facet and fulcrum prominent. Sculpture of similar type to cephalon, if anything still finer.

Discussion. – Boyce (1989) described a similar species, *L. proprium*, from the Lower Member of the Boat Harbour Formation at Eddies Cove West in western Newfoundland. Our material is much more abundant and well preserved and includes a free cheek, which Boyce did not have. The dorsal surface of both cephalon and pygidium in *L. spongiosum* and *L. proprium* is covered with dense pits, which is an important similarity between Newfoundland and Spitsbergen material. The pygidia from both localities are apparently identical, with convex border and seven axial rings of which at least the last is weakly defined. However, cranidia from Spitsbergen differ from the Newfoundland type (and only) cranidium of *L. proprium* in lacking an inflated area in the axial furrow which interrupts its course in front of the occipital ring: this is the 'oblong pre-occipital glabellar lobe' of Boyce's description (Boyce 1989, p. 32). Our cranidia show a uniformly deep axial furrow. Since the cranidia from both localities show a shallow 1S, little more than a smooth area interrupting the sculpture, it seems that the area behind should correspond with L1 glabellar lobe, and the 'oblong lobe' is therefore rather a baccular structure developed in the axial furrow. Ross (1951, pl. 27, fig. 1) figured a generally similar, but ill-preserved pygidium from 'Zone D' in Utah. Landing *et al.* (2012) figured under open nomenclature another similar species from the Rochdale Formation, New York; but it is possible to see that it lacks the surface sculpture of *L. spongiosum*, and the glabellar sides are much more convex.

Family Styginidae Vogdes, 1890

Remarks. – The species described below is one of the earliest known members of superfamily Scutelluoidea, certainly in Laurentia. Possibly, a Swedish species mentioned by Tjernvik (1956) is latest Tremadocian. The long, clavate glabella is clearly similar to that of the upper Cambrian genus *Corynexochus*

and is consistent with the classification of Corynexochida and Scutelluina in the same major clade (Fortey 1990).

Genus *Raymondaspis* Přibyl, 1948

Type species. – *Holometopus limbatus* Angelin, 1854, Sweden.

Raymondaspis? *pingpong* n. sp.

Figure 38A–D, H–K

Holotype. – Cranidium, PMO 208.230 (Fig. 38A–C), Nordporten Member, 106 m.

Material. – Cranidia, PMO 208.282, 223.158; pygidia, PMO 208.210, 208.250–51. *R.*? aff. *pingpong* n. sp. Pygidium, 208.211.

Stratigraphic range. – Lower part of the Nordporten Member, 106 m from base, part of the *Petigurus groenlandicus* fauna, also from the H fauna and from 130 m from base.

Diagnosis. – Referred with caution to *Raymondaspis*, a species with large, gently curved palpebral lobes and a long pygidium without clear border and short axis under half pygidial length.

Etymology. – 'ping pong' – compares the shape of the pygidium with the bat used in that game.

Description. – Small species, with cranidia and pygidia <6 mm long. Cranidium moderately convex, with main downward curvature anteriorly. Glabella clavate, widest across its front margin, where it closely approaches half the maximum cranidial width (tr.) at the posterior margin. The sagittal length of the glabella is close to twice the width across the posterior margin of the occipital ring, which tapers forwards to minimum glabella width at its anterior edge. Thereafter, it expands forwards, hardly at first and progressively more so anteriorly. Axial furrows narrow, but well defined, with slightly outward bulge in pre-occipital region, and with suggestion of a second bulge at about mid-point of palpebral lobes. Furrows terminate in fossulae at anterolateral glabella corners. Pre-glabellar furrow much shallower and gently convex forwards. Occipital furrow transverse, narrow and may be shallow, occipital ring slightly <20% sagittal length and may show faint

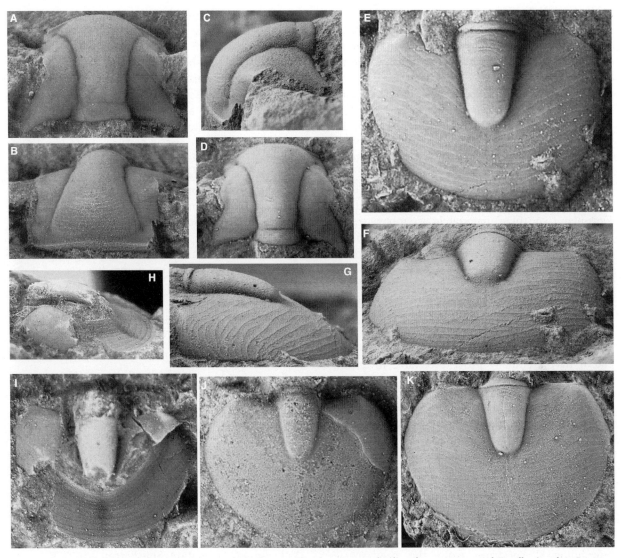

Fig. 38. **A–D, H–K,** *Raymondaspis? pingpong* n. sp. Nordporten Member, *P. groenlandicus* fauna, 106 m, and **H** collection, low *P. groenlandicus* fauna, exact horizon unspecified. **A–C,** Holotype, cranidium PMO 208.230 in dorsal, anterior and lateral views (x8), **H** collection. **D,** Cranidium PMO 208.282 in dorsal and lateral views (x11), **H** collection. **H–I,** Pygidium PMO 208.210 in lateral and dorsal views (x11) prepared to show doublure, 130 m. **J,** Internal mould of pygidium PMO 208.251 in dorsal view (x11), **H** collection. **K,** Pygidium PMO 208.250 in dorsal view (x9), **H** collection. **E–G:** *Raymondaspis* aff. *R.? pingpong.* Pygidium, PMO 208.211 in dorsal and lateral views (x14), 106 m.

median tubercle. Lateral glabellar furrows not seen. Fixed cheeks wide – the posterior limb wider (tr.) at its posterior edge than the narrowest part of the glabella and sloping gently down. Palpebral lobes 25–28% length (exsag.) of glabella (sag.) with very gently curved outline and lacking rims, set far out from glabella (transversely at mid-point 75% of adjacent glabellar width at that section). Steeply downsloping pre-ocular cheeks narrow forwards, because anterior divergence of axial furrows exceeds weak divergence of anterior section of facial sutures. Posterior sections of sutures diverge outwards at 30° to sagittal line with rather straight course to cut posterior margin at high acute angle. Posterior border not defined. Pre-glabellar field lacking medially.

Anterior border is a narrow elevated rim with a forward-facing anterior edge.

Pygidium unusually long, length 75%–80% width, gently and uniformly convex, with the axis standing clearly proud of the pleural fields. Posterior pygidial outline deeply curved – the shape of the whole pygidium making almost three-quarters of a circle. No posterior border. Short, tapering axis, well defined all round, 85% as long as the post-axial field behind it, and 65–75% as wide as long. No axial rings discernable, but half-ring prominent. The front of the pygidial field has an inconspicuous, extremely narrow raised rim, but otherwise the pleural fields are unfurrowed. The downturned facet is deeply cut back extending almost to a level with the end of the axis (presumably the pleural tips were alongside). Doublure concave, with upturned inner margin closely following outline of pygidial margin, extending towards tip of axis, and with a median depression on the mid-line.

Sculpture of quite widely spaced terrace lines, bowed approximately transversely across frontal part of glabella, and on pygidium curving in concave-backwards fashion across post-axial field, and sparse on median part of axis. Internal moulds smooth.

Discussion. – This interesting species is the oldest known member of a group that would last until the Devonian. The type material is assembled from a single bed. However, a second kind of pygidium (Fig. 38E–G), although clearly related, and from a nearby higher locality, cannot be assigned to the same species as *R.? pingpong*. It has a much longer pygidial axis and strong terrace ridges that are widely spaced. It is referred to herein as *R.? aff. pingpong*.

The assignment to *Raymondaspis* is provisional, and it may eventually prove preferable to accommodate the species within a new genus. The early history of styginids was reviewed by Holloway (2007) and is marked by a number of genera with one, or a few species assigned, besides the well-known and much more speciose *Raymondaspis*: these include *Protostygina* Prantl & Přibyl, 1949; *Hallanta* Poulsen, 1965; *Turgicephalus* Fortey, 1975; *Perischoclonus* Whittington, 1963; *Promargo* Holloway, 2007; *Cyrtocybe* Holloway, 2007. *Turgicephalus* was synonymised with *Raymondaspis* by Holloway (2007), although it lies at the opposite end of the morphological spectrum from the new species described here, having both illaenimorph morphology and a transverse pygidium. *Cyrtocybe* is similar to *Turgicephalus* in many ways and indeed was made a sub-genus of *Raymondaspis* (including *Turgicephalus*) by Hansen (2009). While accepting that convexity alone is not a good discriminating character of *Turgicephalus*, as pointed out by Holloway (2007), it remains true that the three Whiterockian species described by Fortey (1980) and Whittington (1965) and placed in *Turgicephalus* form a compact group morphologically. One character that unites them is that on the free cheek, the lateral border furrow shallows posterolaterally where it curves across the cheek, while the inner edge of the genal corner next to it is inflated – the genal spine is lost. Although the type specimen of *Raymondaspis* is not well preserved (see Poulsen 1969), it certainly shows a uniformly deep border furrow, as do several species described by Nielsen (1995) and considered by Holloway (2007) as closely related to the type. The

new species is probably most similar to those *Raymondaspis* species with low cephalic convexity and long pygidia such as *R. vespertina* Ross, 1967, in general glabellar shape and overall convexity. However, *Raymondaspis* spp. have strongly curved palpebral lobes and eyes that describe the greater part of a circle, as well as more transverse pygidia with more or less developed borders. In these characters, the Nordporten species is perhaps more like *Protostygina*. *P. coronula* Adrain & Fortey 1997 from the Tourmakeady Limestone, Ireland, is somewhat similar, but with a more clavate glabella that is more or less effaced anteriorly. Hence, a case might be made for placing *R.? pingpong* into *Protostygina*, and it may lie close to the divergence of several early genera. Rather than add yet another genus to the Styginidae, we prefer to assign the species from the Kirtonryggen Formation with question to *Raymondaspis*, while recognising that a full phylogenetic analysis may change its status in the future.

Order Asaphida, Salter, 1864
Superfamily Asaphoidea Salter, 1864
Family Asaphidae Salter, 1864

Genus *Lachnostoma* Ross, 1951

Type species. – *Lachnostoma latucelsum* Ross, 1951, Garden City Formation.

Other species. – *Paraptychopyge disputa* Fortey, 1975, Valhallfonna Formation, Spitsbergen.

Diagnosis. – Asaphids with long and narrow (tr.) glabella occupying whole width of pre-ocular cranidium, distinctly defined anterior cranidial border, large eyes posteriorly placed; cheeks with or without genal spine. Hypostome with very wide flat borders, with or without marginal spines, wide anteriorly, and shallow posteromedian notch with inverted u-shaped profile, middle body well defined, tapering backwards; pygidium with long and narrow axis, wide pleural fields and faintly defined border; doublure broad and may curve inwards towards axis (emended from Ross 1951).

Discussion. – Asaphids are difficult to classify, and the species under consideration below is no exception. In assigning the species to *Lachnostoma*, an important feature was considered to be the hypostome. Few asaphids have lateral hypostomal borders that are wide anteriorly – most widen backwards towards the median fork, while those lacking a fork have narrow borders. The hypostomal border of the new species is wide at the front – well in front of the maculae. One comparable genus in this regard is *Ptyocephalus*, the hypostome of which has been described from several species (Ross 1951; Hintze 1953; Fortey 1975). The narrow (tr.) and long cranidium of the new species is also generally like that of *Ptyocephalus*. However, it would scarcely be possible to place our new material in *Ptyocephalus* because species in that genus are united in having several distinctive characters lacking in our material. For instance, there is a strong vincular groove on the cephalic doublure of *P. declevita* (Ross 1951, pl. 22, fig. 8), and the free cheeks of *Ptyocephalus* are curiously truncate laterally, while the pygidium has a pentagonal form unusual in the Asaphidae as a whole. On balance, *Lachnostoma* Ross, 1951, provides a better comparison. Although the hypostome of the type species has a more elaborate margin, carrying auxiliary marginal spines, it is generally similar to the species below in having a narrow notch with an inverted u-shaped profile and wide borders. Pygidial morphology is also very similar, although the cranidium of *L. latucelsum* has more divergent anterior sections of the facial sutures resulting in wider pre-ocular fixed cheeks, as well as smaller eyes. We also considered *Stegnopsis* Whittington, 1965, which has cranidial morphology like that of *Lachnostoma*, but an *Isotelus*-like hypostome, which in our opinion rules out a close relationship both to our material and to *Lachnostoma*. Fortey (1975) described a species from the basal Valhallfonna Formation as *Paraptychopgye disputa*, but without knowledge of its hypostome. It is now obvious that this species is very close to the new species below. Both *P. disputa* and the new species are better referred to *Lachnostoma*. This does require modification of the generic diagnosis to include more variation in the sutural outline, eye size and width of doublure.

Lachnostoma platypyga n. sp.

Figure 39

Holotype. – Cranidium, CAMSM X 50188.2 (Fig. 39G); top 3 m of Nordporten Member.

Material. – Cranidia, PMO 208.167, 208.297(2), CAMSM X 50188.3–5, X 50188.86 (juv); free cheeks, PMO 208.164, CAMSM X 50188.7, X 50188.10; pygidia, PMO NF 2556, CAMSM X 50188.8–9; hypostome, PMO 223.244(1), CAMSM X 50188.11; growth stages, PMO 208.110, 208.145–6, 208.165, 208.297(1).

Stratigraphic range. – Top 3 m of Nordporten Member of Kirtonryggen Formation, *Lachnostoma platypyga* fauna (Blackhillsian).

Etymology. – Greek: 'flat tail'.

Description. – Cranidium almost twice as long as width across frontal glabellar lobe. Glabella narrowest medially, expanding in width in front of eyes such that frontal lobe occupies all anterior width of cranidium. Axial furrows shallow and diffuse, faintly indicating 'waisting' of mid-part of glabella, and occipital ring not defined. No muscle impressions clearly shown except for a pair of circular impressions in front of anterior end of palpebral lobes. Glabellar tubercle lies at one-quarter glabellar length. Frontal lobe slopes downwards to abruptly defined anterior cranidial border one-sixth of glabellar length (sag.), which is gently concave. Palpebral lobes flat and large, one-third length of cranidium and highly curved, with rather weakly indicated palpebral rims. Post-ocular fixed cheeks are narrow (exsag.) and strap like, posterior section of facial suture turning outwards at a right angle to sag.line. Anterior sections, by contrast, diverge outwards at a rather low angle in front of the palpebral lobes and then progressively inwards anteriorly in a rather smooth arc and meet on mid-line without evidence of a median point. A series of small cranidia have relatively larger palpebral lobes than fully developed individuals, usually a more prominent glabellar tubercle, and the anterior border is longer.

Free cheeks wide, approximately as long (exsag.) as wide (tr.), and lacking genal spine, the genal angle being obtusely rounded. Paradoublural line shows on dorsal surface passing close to eye. Border less distinct than on cranidium. Eye strongly curved, not elevated on socle, lenses minute. Wide doublure underlies most of cheek and curves back close to dorsal surface, carrying 20–25 parallel terrace lines that become faint or absent posteriorly. Small panderian opening well in from genal angle and far removed from inner edge of doublure. No clear vincular groove.

Hypostome is wider than long, with large, flat lateral borders having maximum width in front of maculae. Middle body tapers posteriorly evenly and is well defined as far as the maculae, which are deep pits indenting the flat lateral borders. Posterior margin of middle body close to inner edge of fork and marked by a transverse furrow. Posteromedian fork is shallow, wide U shape. Terrace ridges more or less transverse on border, lacking on middle body. Small hypostome shows suggestion of marginal angulations comparable with spinose margin of the type species.

Pygidium 60% as long as wide, of low convexity. Pygidia are the commonest sclerites and may attain more than 5 cm width, the largest examples also being the least convex. There is a somewhat ill-defined, slightly concave to flattened border, most marked behind the axis. The narrow axis occupies about one-fifth of pygidial width at anterior end, or slightly more on smaller specimens, and is weakly funnel shaped, extending to two-thirds of pygidial length. Axial furrows weak. Three faint axial rings are visible on internal moulds, and about twice this number of muscle impressions on the dorsal surface. Two or three pairs of weak ribs are present on some specimens: these fade away rapidly from the axis. Doublure very wide, recurved back dorsally almost to axis, hence inner margin convex forwards, curving outwards again anteriorly to underlie facet. About twenty terrace lines that follow inner margin of doublure and posterior pygidial margin, and a compromise in between.

Sculpture lacking on cranidium and weak terraces on free cheek; very dense fine terrace ridges on pygidial pleural fields more or less concentric and parallel to pygidial margin.

Small cranidia have relatively large glabellar tubercle and larger palpebral lobes, reflected on immature free cheeks as larger eye. In both, the border is sharply defined. Transitory pygidium also with sharply defined border and notable posteromedian arch, much like those of *Lachnostoma latucelsum*.

Discussion. – Fortey (1975) assigned a species from a short stratigraphic distance above the present one in the basal part of the Valhallfonna Formation to *Paraptychopyge* Balashova, 1964. This species, *P. disputa*, is very much like the new one described above and must be congeneric. The resemblance of both Spitsbergen species to ptychopgyinids is now regarded as misleading, being based particularly on the long narrow glabella and very wide pygidial doublure. Discovery of the hypostome of the new species shows that it is completely unlike that of the many ptychopyginids for which they are known (e.g. Balashova 1976, pl. 14) – the latter have narrow borders and a short deep fork. We now consider that the Spitsbergen forms are not closely related to the Baltic taxa and that any resemblance they share is because of convergence. *Lachnostoma* is the only genus that passes from the Nordporten Member into the Valhallfonna Formation, where there is a strong facies change. *Lachnostoma platypyga* is so similar to *L. disputa* that we considered the possibility that the differences between them might be due to intraspecific variation. However, the cranidial border on *L. disputa* is about half as wide as that on *L. platypyga*, and the pygidium is proportionately wider; a large example of *L. disputa* shows an embayed pygidial margin, and the doublure does not curve so strongly forwards; there are enough examples to be sure of the consistency of this difference. It is reflected in the course of the terrace ridges on the pygidial doublure, which are all nearly parallel on *L. disputa*, whereas the inner ones on *L. platypyga* follow a different course to those near the pygidial margin. We conclude that there are two distinct species above and below the base of the Valhallfonna Formation. Both are readily distinguished from the slightly younger type species from the Great Basin in having much less divergent anterior sections of the facial sutures, wider doublures, and (in *L. platypyga*) a mature hypostome lacking additional marginal spines.

Genus *Stenorhachis* Hintze & Jaanusson, 1956

Type species. – *Isoteloides genalticurvatus* Hintze, 1953, from Utah, original designation.

Stenorhachis n. sp. A

Figure 40B–D

Material. – Pygidia, PMO 208.160–3.

Fig. 39. Lachnostoma platypyga n. sp. Topmost 3 m Nordporten Member, *L. platypyga* fauna. **A**, Free cheek CAMSM X 50188.10 in dorsal view (x2) to show doublure and panderian opening. **B**, Cranidium CAMSM X 50188.86 in dorsal view (x9), immature. **C–D**, Small cranidium PMO 208.297 (2) in dorsal and anterior views (x18). **E**, Large free cheek CAMSM X 50188.7 in dorsal view (x1). **F**, Immature free cheek PMO 208.164 in dorsal view (x8). **G**, Holotype, larger cranidium CAMSM X 50188.2 in dorsal view (x3). **H**, Cranidium CAMSM X 50188.3 in dorsal view (x4). **I–J**, Latex cast of incomplete mature cranidium CAMSM X 50188.4 in dorsal and lateral views (x4).). **K, M–N**, Hypostome, **K**, latex cast prepared from counterpart of specimen **M** CAMSM X 50188.11, incomplete but showing wide border, in ventral view (x2), and (**N**) enlargement showing sculpture. **L**, Smaller hypostome PMO 223.244 (1) in dorsal view (x12). **O–Q**, Pygidium prepared to show doublure PMO NF 2556 in dorsal, posterior and lateral views (x1.5). **R–S**, Transitory pygidium PMO 208.297 (1) in dorsal and posterior views (x20). **T**, Latex cast from pygidium CAMSM X 50188.8 in dorsal view to show exterior surface (x2). **U**, Immature pygidium PMO 208.110 in dorsal view (x6) showing relatively wide axis and defined border at small size. **V**, Small pygidium PMO 208.146 in dorsal view (x5). **W**, Pygidium PMO 208.165 in dorsal view (x3).

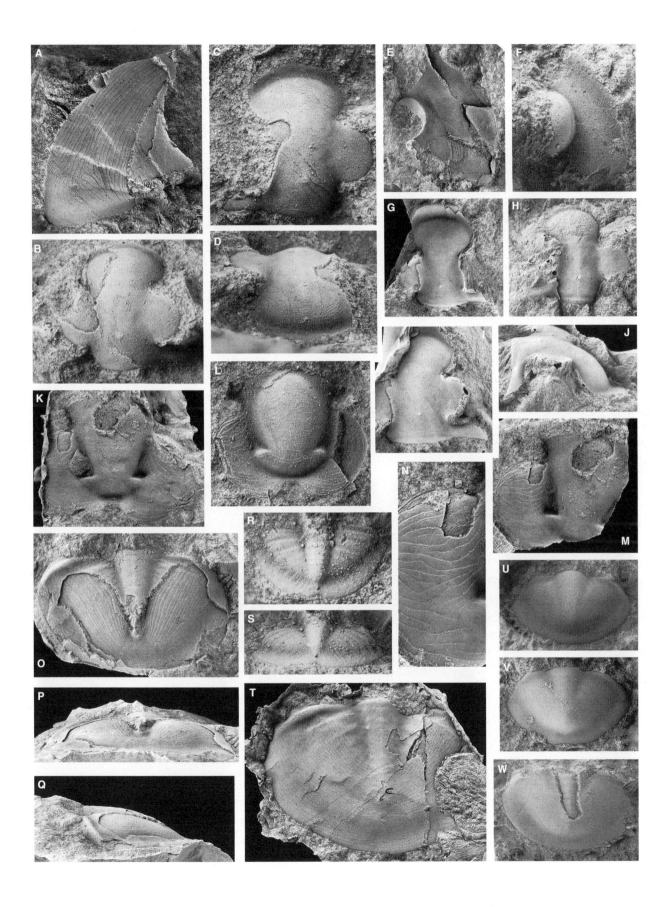

Stratigraphic range. – Topmost 3 m of the Nord-porten Member, *Lachnostoma platypyga* fauna, Blackhillsian (early Floian).

Discussion. – One bed at the top of the Nordporten Member is crowded with remarkable and large pygidia of what is undoubtedly a new asaphid species. Unfortunately, this is all we know about it, as no other sclerites have been certainly identified. It has a spade-like profile, extremely low convexity, exceptionally short axis and very wide doublure that is reflexed back immediately under the dorsal surface. The inner margin of the doublure almost reaches the axis on all sides. The dorsal surface sculpture of terrace ridges and those on the doub-lure are also distinctive. It must remain under open nomenclature until its cephalic parts are discovered. The assignment to *Stenorhachis* may be dis-proved when that happens. However, the type and only species of that genus from Utah is only slightly younger and occurs in the same fauna as abundant *Lachnostoma*. It, too, is of extremely low convexity (sag., tr.), and the pygidium does have an unusually broad doublure, although the axis is longer and the overall shape is quite different from the Nordporten species. *Stenorhachis* seems the most plausible genus in which to place the Spits-bergen species. A small cranidium has been discov-ered which does not belong to the commoner

species *Licnocephala platypyga* (Fig. 40A), and it might be assumed therefore that it is that of *Stenorhachis* sp. A., but there is no particular reason to find this plausible.

Family Remopleurididae Hawle and Corda, 1847

Genus *Eorobergia* Cooper, 1953

Type species. – *Robergia marginalis* Raymond, 1925, Edinburg Formation, Virginia.

Eorobergia n. sp. A

Figure 40F

Material. – Cranidium, PMO 208.168.

Stratigraphic range. – Nordporten Member, 1 m from top, *Lachnostoma platypyga* fauna (Blackhill-sian, early Floian).

Description. – Cranidium only known, though it is distinctive; flat in posterior part, with very broad anterior tongue, downsloping at front. Total dorsal length 90% width across palpebral

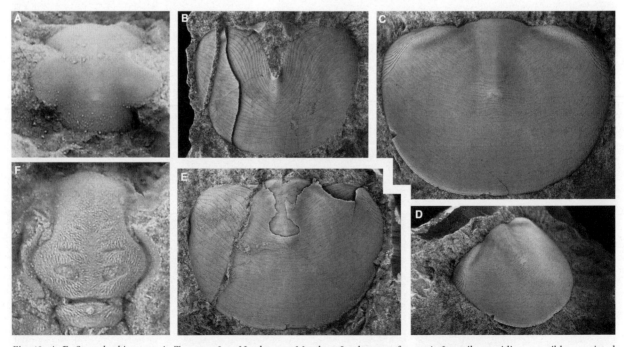

Fig. 40. A–E, *Stenorhachis* n. sp. **A.** Topmost 3 m Nordporten Member, *L. platypyga* fauna. **A,** Juvenile cranidium possibly associated PMO 208.297(3) in dorsal view (x17). **B,** Pygidium prepared to show extraordinarily wide doublure PMO 208.160 in dorsal view (x1.25). **C–D,** Pygidium PMO 208.162 in dorsal (x3) and oblique lateral (x2) views. **E,** Pygidium PMO 208.161 in dorsal view (x1.5). **F,** *Eorober-gia* n. sp. A. Top 1 m Nordporten Member, *L. platypyga* fauna. Cranidium PMO 208.168 in dorsal view (x15).

lobes; half pre-occipital length is accounted for by the anterior tongue. Occipital ring is one-sixth of the total dorsal length. Glabella of typical remopleuridioid form, expanding to occupy the space between palpebral lobes in front of the occipital furrow and then contracting again forwards, back to about the same width (tr.) as the occipital ring (tr.). Wide tongue hardly expands in width forwards until terminated at narrow (exsag.), horizontal anterior border, incompletely preserved. Glabellar furrows prominent as depressed areas within glabella, S1 wide (exsag), with almond shaped outline – curved anterior edge, straight inward sloping posterior edge; S2 close to transverse, narrow, hardly curved; S3 short, less than half length of S2, opposite anterior ends of palpebral lobes. Latter are narrowly lunate, flat, of rather even width (tr.) along their length, smooth, defined by palpebral furrows that are shallower medially. Occipital ring well defined, by furrow medially gently curving forwards, widest at mid-width, with very prominent median tubercle advanced from mid-point, carrying obscure dorsal pits. Whole surface of glabella and occipital ring (apart from along mid-line) covered with distinctive, slightly undulating, coarse raised ridges arranged more transversely medially, more exsagitally laterally and breaking up anteriorly.

Discussion. – Sadly, we only know this distinctive remopleuridid – the only one from the Kirtonryggen Formation – from a single, well-preserved cranidium, which precludes naming it formally. The very broad and long (sag.) anterior tongue, coupled with the rugulose surface sculpture, serves to distinguish the species from all others of approximately the same age. If the pygidium were known, it might perhaps be preferably be assigned to *Menoparia* Ross, 1951, and the generic assignment is regarded as provisional, although the narrow cranidial border is consistent with *Eorobergia*. Along with *Conophrys*, this species is indicative of a short-lived deepening event at the very top of the Kirtonryggen Formation, where species belonging to families more widespread than Bathyuridae briefly appear.

Superfamily Cyclopygoidea Raymond, 1925
Family Symphysurinidae Kobayashi, 1955

Discussion. – Fortey (1983) noted that a species of *Symphysurina* from the Lower Ordovician of Newfoundland had nine thoracic segments, a feature which separated it from all known members of the Asaphidae, to which the genus had been assigned. The resemblance of this genus to asaphids was a matter of convergence, which is common in effaced trilobites. There are several features that suggest a possible relationship of *Symphysurina* to Cyclopygoidea: notably, the presence in some species of *Symphysurina*, including the type species, of a series of vincular notches in the genal doublure, and the articulation of the anterior thoracic segment very close to the axis, which is invariable in Cyclopygidae, and almost so in Nileidae. The presence of a ventral median suture is neither here nor there because it is present too in primitive nileids (*Platypeltoides*) and cyclopygids (*Prospectatrix*). However, it is uncertain where *Symphysurina* would fit into this larger group, and for the moment, it is preferable to recognise it in a separate family, which would include *Symphysurina* and *Eurysymphysurina* n. gen. and, possibly, *Randaynia* Boyce, 1989.

Genus *Eurysymphysurina* n. gen.

Type species. – *Eurysymphysurina spora* n. sp.

Diagnosis. – Symphysurinids with wide, effaced cephalic shield differing from *Symphysurina* in having a relatively small pygidium, with no posterior spine, and steeply sloping posterior pygidial border, on to which the axis may extend. Cephalon transverse, *Illaenus* like, almost entirely without furrows, highly convex (sag.), with very large triangular, blade-like cheeks with faint border.

Discussion. – Symphysurinids span the Cambrian–Ordovician boundary. It is becoming clear that they embrace a considerable range of morphology and, rather as has been the case with the contemporary *Hystricurus*, there are several discrete clades within the group. The species from the Spora Member of the Kirtonryggen Formation is very far from the type species of *Symphysurina*, *S. woosteri* Walcott, 1924, and indeed most of the Early Ordovician members of the genus. *Symphysurina woosteri* has a relatively well-defined glabella, genal spines well differentiated from the genal field, and pygidium with a distinct axis and posterior spine; the Spitsbergen species has none of these attributes. The revision of the whole group may result in the erection of several genera, which is beyond the scope of this work. However, it seems inconceivable that the distinctive species described below would finish up in the same clade as the type species in any future analysis, and it is appropriate to split off the new taxon herein, which is younger and more derived than others in the group. Interestingly, the species is convergent, not only on *Illaenus*, but also on the bathyurid *Uromystrum affine* described above. These illaenimorphs were presumably adapted to very similar circumstances in the epeiric environments of the Early Ordovician carbonate platform. We also provisionally assign *Symphysurina elegans* Poulsen, 1937, to the same genus, which shows a similarly tumid glabella, although it is less effaced than the type species of *Eurysymphysurina*.

Etymology. – Greek: '*eury*' – meaning 'wide' + *Symphysurina*.

Eurysymphysurina spora n. sp.

Figure 41

Holotype. – Cranidium, PMO 208.194 (Fig. 41C–E), Spora Member.

Material. – Cranidia, PMO 208.064, 208.193, 208.194; free cheeks, PMO 208.067, 208.195, PMO 221.250–1; pygidia, PMO 208.052, CAM X 50188.62.

Stratigraphic range. – Spora Member (undivided) of the Kirtonryggen Formation, Early Ordovician, Stairsian (Tremadocian), *Svalbardicurus delicatus* fauna.

Diagnosis. – *Eurysymphysurina* species with widely triangular free cheeks, lacking any surface sculpture apart from raised ridges on the posterior pygidial margin.

Etymology. – 'spora' – from its occurrence in the Spora Member.

Description. – Free cheeks probably grew to at least 2 cm length, so this is not a small trilobite, although the cuticle is relatively thin and exfoliates readily. Cranidium is exceeding convex, such that its length (sag.) in dorsal view is exceeded by its height in anterior view, and it has a recurved profile in lateral view. With the free cheeks in position, the entire cephalon would have been crescentic in outline and very transverse (about three times as wide across the genal spines as long sagittally) with the genal areas swept backwards alongside the anterior part of the thorax. The cranidium is generally featureless, 75%

as long as wide in dorsal view, with convex (tr.) glabella occupying much of it, but indistinctly defined except at the posterior margin, where it is about two-thirds cranidial width. Axial furrows are not defined, and it is not possible to distinguish the glabella anteriorly. However, the front of the cranidium is gently bowed outwards in a narrow sloping border, and presumably the glabella terminates anterior to this. Narrow (tr.) triangular post-ocular fixed cheeks, with no border, defined by straight sutures that diverge 30–35° from sag. line. Pre-ocular sutures diverge at a similar angle and curve in a smooth arc around anterolateral cranidial corners before running along cranidial edge which is bowed outwards around mid-line. Palpebral lobes gently curved, sloping forwards in dorsal aspect, about one-third cranidial length (exag.), though foreshortening in this view accentuates their relative length. Sculpture lacking.

Free cheek in dorsal aspect narrowly triangular and curved backwards, the bulk of the genal field facing outwards and sloping steeply. In lateral view, the eye forms the high point, and the cheek narrows gradually and continuously away from it to terminate in an acute genal spine. The slightly concave lateral border is apparent anteromedially but narrows distally fading out about half way along genal margin. No sign of a posterior border. Eye lobe is low and gently curved, set on the cheek directly; the facial suture curves down immediately in front of it. Doublure is reflexed back inside genal corner, forming an anterior tubular structure.

Pygidium relatively small and twice as wide as long (dorsal view), with a very steep posterior 'wall', which is divided from the rest of the pleural field by an abrupt change in slope. The fact that the axis carries on to this part of the pygidium indicates that this apparent border is a modified part of the pleural field. Poorly defined axis itself is widest at the prominent articulating half-ring where it is little wider (tr.) than the adjacent pleural area, tapers gently to an obscure posterior termination, the first two axial rings (which are three to four times as long as wide) relatively clearly defined with a suggestion of a third and possibly a fourth ring. Pleural field unfurrowed. Posterior margin with a median embayment. Unlike the rest of the exoskeleton, the sloping posterior margin carries a sculpture of raised lines. It seems plausible that the modifications on the pygidial margin are coaptative, with the sloping 'wall' tucking under the cephalic brim during enrolment. If so, it would imply spiral enrolment style, which is unknown in Asaphidae.

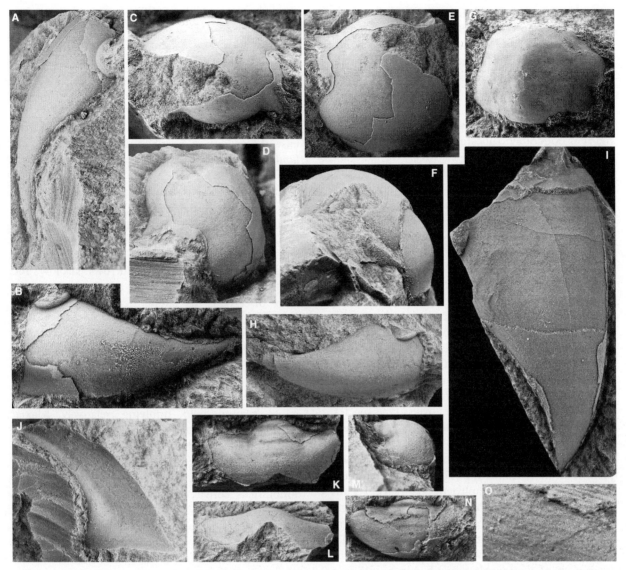

Fig. 41. *Eurysymphysurina spora* n. sp. Spora Member, *S. delicatus* fauna (1–20 m, undivided horizons). **A–B**, Free cheek PMO 208.195 in dorsal and lateral views (x5). **C–E**, Holotype partly exfoliated cranidium PMO 208.194 in dorsal, oblique lateral and anterior views (x6). **F**, Cranidium PMO.208.064 in lateral view (x4). **G**, Incomplete cranidium but showing palpebral lobe on right PMO 208.193 in dorsal view (x3). **H**, Free cheek PMO 221.251 in lateral view (x5). **I**, Large free cheek with shorter spine PMO 208.067 in plan view (x4). **J**, Doublure of free cheek PMO 221.250 in dorsal view (x4). **K–M**, Incomplete pygidium PMO 208.052 in dorsal, posterior and lateral views (x4). **N–O**, Pygidium CAMSM X 50188.62 in dorsal view (x4), and (**O**) enlargement of small part of border showing the sculpture of dense raised lines.

Discussion. – No described species of *Symphysurina* is similar to the one under consideration, with its extended triangular cheeks and small pygidium. The type species of *Symphysurina* has a large pygidium carrying a posteromedian spine, a well-defined glabella, and small free cheeks.

Genus *Randaynia* Boyce, 1989

Type species. – *Randaynia saundersi* Boyce, 1989, from the Boat Harbour Formation, western Newfoundland, original designation.

Discussion. – Two species from western Newfoundland were described by Boyce (1989) in *Randaynia*, who reserved judgement about assigning the genus to a family. Loch (2007, p. 70) gave reasons for including the genus in the bathyurid subfamily Bathyurellinae and added an additional species from Oklahoma. When redescribing *R. taurifrons* (Dwight, 1884) from the Rochdale Formation, Landing *et al.* (2012) also assigned *Randaynia* to Bathyuridae. Loch (2007) extended the compass of *Randaynia* to include *Bathyurellus affinis* Poulsen, 1937, a species now known from a good deal of material in the Kirtonryggen Formation (above).

The assignment of *Randaynia* to Bathyuridae is not, in our view, appropriate. Effacement is a polyphyletic character and does not, by itself, form a sound basis for definition of genera. Although generally effaced, it is possible to make out the glabellar outline of both *Randaynia saundersi* and *R. langdoni* Boyce, 1989. The glabella is sub-rectangular and occupies much of the medial part of the cranidium, and it is noticeably truncate anteriorly. On '*B.*' *affinis* by contrast, the glabella is smaller, elongate and pointed anteriorly – essentially that of a typical bathyurelline that has become effaced. Although the pygidium is also effaced, internal moulds show faint indications of the usual four segments expected in a bathyurid. Hence, '*B.*' *affinis* is accommodated in the bathyurelline genus *Uromystrum* herein. The pygidium of *R. langdoni* shows up to seven pairs of muscle attachment areas (Boyce 1989, pl. 38, fig. 10), which is not a feature of bathyurids; with the exception of *Catochia*, bathyurids have four segments in the pygidium, whether they are bathyurines or bathyurellines, and this is even maintained in effaced genera, such as *Benthamaspis*. Boyce (1989) proved the presence of six or seven thoracic segments in *R. saundersi,* whereas other bathyurids have nine or ten. Furthermore, Boyce's type species, *R. saundersi,* is part of the Stairsian, late 'hystricurid' fauna (Boyce, 1989, p. 10), and predates the bathyurid radiation that happened in the subsequent Tulean–Blackhillsian. The basal, plesiomorphic and probably paraphyletic bathyurid genus *Peltabellia* is younger than *Randaynia* at its first appearance. Putting this evidence together, it seems that Boyce's (1989) caution was correct and that *Randaynia* is an unlikely bathyurid. However, the description of *Eurysymphysurina* herein points to a species with a free cheek much like that of *Randaynia saundersi* that is not a bathyurid. It seems plausible to us that *Randaynia* is another member of the Family Symphysurinidae that radiated in the earlier Tremadocian *before* the bathyurids expanded in the later Tremadocian–Floian. Hence, we place *Randaynia* close to *Eurysymphysurina* here and exclude it from Family Bathyuridae.

Randaynia n. sp. A

Figure 42A–D, ?E

Material. – Pygidium, PMO 208.209.

Stratigraphic range. – Known from 106 m from base of Nordporten Member, upper *Petigurus groenlandicus* fauna.

Description. – Evenly convex and generally effaced pygidium two-thirds as long as wide. Axis not greatly vaulted transversely, widest anteriorly (where about 0.4 transverse pygidial width) and tapering gently to about 0.8 pygidial length to a broadly rounded tip. Axial furrows shallow and faint, as are axial rings, with two or three broad rings discernable as lateral depressions, of which only the first passes across the mid-part of the axis. The prominent anterior half-ring, however, is deeply defined on the internal surface. One pair of pleural furrows is deepest behind the inner portion of facet, fading both adaxially and distally. Steeply downturned facet extends a long way across pleural field implying relatively adaxial articulation in posterior part of the thorax. In contrast to the rather routine dorsal surface, the doublure is peculiar, and we are not aware of similar morphology being described. Posteriorly, the doublure is convex downwards, producing a ventral, bladder like bulge, but laterally becomes reflexed dorsally close to the underside of the lateral part of the pygidial pleural field. We can find no evidence of surface sculpture on adherent cuticle fragments.

Discussion. – Despite its lack of a border, we regard this as a *Randaynia* species. The large axis is characteristic. It is also worth noting the differences from bathyurid pygidia, since it might be thought to somewhat resemble, for example, the pygidium of *Benthamaspis*. The doublure of bathyurids is reflexed at first more or less horizontally before

---

Fig. 42. **A–D, ?E,** *Randaynia* n. sp. A. Nordporten Member, *Petigurus groenlandicus* fauna, 106 m. **A–D,** Pygidium PMO 208.209 in dorsal (x5), lateral, dorsoposterior and posterior (x4) views, to show inflated pygidial doublure. **E,** Very tentatively associated cranidium PMO 223.249 in dorsal view (x4), 105 m. **F–I,** *Pilekia* sp. cf. *P. trio* Ross, 1951. Spora Member, *S. delicatus* fauna (1–20 m, undivided horizons). **F–G,** Cranidium PMO 208.053 in dorsal and anterior views (x5). **H,** Enlargement of **B** showing sculpture. **I,** Incomplete cranidium PMO 208.054 in dorsal view (x6). **J–N,** *Conophrys* sp. 1. Topmost 3 m Nordporten Member, *L. platypyga* fauna. **J–L,** Cranidium PMO 223.243 in dorsal, anterior and lateral views. **M–N,** Cranidium PMO 223.244 (2) in dorsal and anterior view (x22). **O,** Gen. et sp. indet. 1. Spora Member, *S. delicatus* fauna (1–20 m, undivided horizons). Cranidium PMO 208.187 in dorsal view (x8). **Q–U,** Gen. et sp. indet. 2. Topmost 2 m Bassisletta Member. **Q–S,** Pygidium PMO 208.215 in dorsal, posterior and lateral views (x3), and (**T**) enlargement showing edge of doublure. **U,** Cast of incomplete pygidium PMO 223.247 in dorsal view (x2). **P,** Unknown cranidial fragment from Spora Member, *S. delicatus* fauna (not described) PMO 208.050 in dorsal view (x2).

Incertae Ordinis
Family Shumardiidae Lake, 1907

Genus *Conophrys* Callaway, 1877

Type species. – *Conophrys salopiensis* Callaway, 1877, Tremadocian, Shropshire, UK, revised by Fortey & Owens (1991).

Discussion. – The most recent and comprehensive phylogenetic revision of shumardiids was by Waisfeld *et al.* (2001). These authors recognised *Conophrys* as a 'grade group' – a paraphyletic assemblage from which other shumardiid genera were probably derived. The type species is from the late Tremadocian of England, and therefore, only slightly older than a species from the top of the Norporten Member (early Floian).

Conophrys sp. 1

Figure 42J–N

Material. – Cranidia, PMO 223.243–4(2).

Stratigraphic range. – Top 3 m of Nordporten Member, *Lachnostoma platypyga* fauna (Blackhillsian, early Floian).

Discussion. – The appearance of the widespread genus *Conophrys* at the top of the Kirtonryggen Formation is consistent with a change in facies beneath the Valhallfonna Formation. Pygidia of shumardiids like *Conophrys* have proved informative taxonomically (Fortey & Owens 1991; Ebbestad 1999; Waisfeld *et al.* 2001). Without an associated pygidium, therefore, the species from the top of the Kirtonryggen Formation cannot be named formally. With its well-defined glabella with broad lateral lobes, it does conform to *Conophrys*, though with narrow (tr.) cheeks compared with Tremadocian *S. pusilla* (Sars) (see Ebbestad 1999) and *S. salopiensis*. It has a sculpture of scattered pits. Early Floian (= Arenig) shumardiids are little known. *Conophrys nericiensis* (Wiman, 1905) is of similar age, and its Swedish type material was refigured by Hoel (1999), along with some new specimens from the Oslo region. Like *C.* sp. 1, all this material has narrow (tr.) cheeks. The front of the glabella is visible on the Swedish material, but the lateral lobes are smaller than the Nordporten Member examples, and the front out-

line of the glabella differs. The front of the glabella on the new specimens of Hoel (1999, figs 3A–B) is noticeably effaced and broader. *Conophrys* sp. 1 is very likely a new species, but open nomenclature is preferred here pending new discoveries.

Incertae sedis

Gen. et sp. indet 1

Figure 42O

Material. – Cranidium, PMO 208.187.

Stratigraphic range. – Spora Member (undivided), *Svalbardicurus delicatus* fauna, Tremadocian (Stairsian).

Discussion. – A partial cranidium displays a pyriform glabella with a posterolateral, depressed lobe; long, downsloping pre-glabellar field extending into a broad anterolateral fixigenal area, and with a narrow, rim-like cranidial border. The base of a strong eye ridge is preserved but not the palpebral lobe. There is a distinctive surface sculpture of big, close tubercles along the mid-line of the glabella, but the pre-glabellar area shows only fine caeca. This is a distinctive fragment that we cannot match in Early Ordovician faunas. *Hillyardina levis* Boyce, 1989, has a similarly long pre-glabellar area, but Boyce's specimens include examples with both defined and effaced glabellar furrows and may be a mix of two taxa. In any case, none of them show tuberculate sculpture. Our specimen is figured because it is distinctive enough to be recognised elsewhere.

Gen. et sp. indet. 2

Figure 42Q–U

Material. – Pygidia, 208.215, 223.247.

Stratigraphic range. – Top 2 m Bassisletta Member, upper *Chapmanopyge* fauna.

Description. – This distinctive, convex pygidium lacking border has a very short axis, faintly, but sufficiently, defined to see its truncate termination. At one point, we considered that the specimens were incomplete at the front, but preparation of the peculiar abbreviated (tr.) facet shows this not to be the case. This species evidently grew to a large size since we have a fragment 5 cm across. No evidence of dorsal sculpture has been seen.

Discussion. – This species has puzzling morphology that does not permit its ready classification. We regard it as conceivable that it is the pygidium associated with 'Bathyurine n. gen. et sp. A' (p. 54) because the convexity of the axis and its width would be compatible, but in other respects, the axial morphology does not suggest a bathyurid at all, and there is no real evidence that they belong together. Hence, a tentative name is the preferred option at the moment.

Proetid metaprotaspis

Figure 33L

Material. – One specimen, PMO 223.242.

Stratigraphic range. – Uppermost part of Nordporten Member, *Lachnostoma platypyga* Fauna.

Discussion. – Little evidence of early growth stages has been recovered from the Kirtonryggen Formation, so this well-preserved metaprotaspis is of interest. It obviously conforms to the 'adult-like' form characteristic of the Order Proetida. With its prominent, segment-related fixigenal tubercles, it is likely to belong to the aulacopleuroid–bathyurid clade. There is no real evidence to link it with any particular species in the upper Kirtonryggen fauna. *Ischyrotoma* n. sp. A is from the same strata, but since bathyurids like *Petigurus nero* are found a few metres below in the section, there are other possibilities. It is figured with *Psalikilopsis* purely for convenience.

Acknowledgements. – We most particularly thank Mrs Di Clements for help in organising the material, arranging loans and preparing labels. Cyrille Delmer is gratefully acknowledged for his help with the diagrams and figures. Phil Crabb and Phil Hurst did most of the photography. Staff at the Sedgwick Museum, Cambridge, Yale Peabody Museum and U.S. National Museum kindly made specimens available for study. Lucy McCobb (Cardiff) was generous enough to allow us to use photographs she had made of Poulsen's types. Dr M. P. Smith is thanked particularly for processing limestone samples and identifying some critical conodont species he recovered. Much of the preparation was done by RAF, the rest by DLB who also provided some photographs. RAF's work was completed while in receipt of a Leverhulme Emeritus Fellowship, which is gratefully recognised here.

We both acknowledge the support in the field from the late Gunnar Henningsmoen, the late Aage Jensen, Leif Koch and Kjetil Gram. DLB acknowledges field support from the former Norges Almenvitenskapelige Forskningsråd (NAVF) and the University of Oslo.

Useful referee comments were provided by Dr S. R. Westrop and Dr J. F. Taylor.

References

Adrain, J.M. 2011: Class Trilobita Walch, 1771. *In* Zhang, Z.-G. (ed.): *Animal Biodiversity: An Outline of Higher Level Classification and Survey of Taxonomic Richness.* Zootaxa, *3148*, 104–109.

Adrain, J.M. & Fortey, R.A. 1997: Ordovician trilobites from the Tourmakeady Limestone, western Ireland. *Bulletin of the British Museum (Natural History) Geology series 53*, 79–115.

Adrain, J.M., McAdams, N.E.B. & Karim, T.S. 2012: The Lower Ordovician (upper Floian) bathyurid trilobite *Aponileus* Hu, with species from Utah, Texas, and Greenland. *Zootaxa 3293*, 1–67.

Adrain, J.M. & Westrop, S.R. 2005: Lower Ordovician trilobites from the Baumann Fiord Formation, Ellesmere Island, Arctic Canada. *Canadian Journal of Earth Sciences 42*, 1523–1546.

Adrain, J.M., Westrop, S.R., Landing, E. & Fortey, R.A. 2001: Systematics of the Ordovician trilobites *Ischyrotoma* and *Dimeropygiella*, with species from the type Ibexian area, western USA. *Journal of Paleontology 75*, 947–971.

Adrain, J.M., Lee, D.-C., Westrop, S.R., Chatterton, B.D.E. & Landing, E. 2003: Classification of the trilobite subfamilies Hystricurinae and Hintzecurinae subfam. nov., with new genera from the Lower Ordovician (Ibexian) of Idaho and Utah. *Memoirs of the Queensland Museum 48*, 553–586.

Adrain, J.M., McAdams, N.E.B. & Westrop, S.R. 2009: Trilobite biostratigraphy and revised bases of the Tulean and Blackhillsian Stages of the Ibexian Series, Lower Ordovician, western United States. *Memoirs of the Australasian Association of Palaeontologists 37*, 541–610.

Adrain, J.M., McAdams, N.E.B. & Westrop, S.R. 2011a: Affinities of the Lower Ordovician (Tulean; lower Floian) trilobite *Gladiatoria*, with species from the Great Basin, western United States. *Memoirs of the Association of Australasian Palaeontologists 42*, 321–336.

Adrain, J.M., McAdams, N.E.B., Westrop, S.R. & Karim, T.S. 2011b: Systematics and affinity of the Lower Ordovician (Tulean; Lower Floian) trilobite *Psalikilopsis*. *Memoirs of the Association of Australasian Palaeontologists 42*, 369–416.

Archer, J.B. & Fortey, R.A. 1974: Ordovician graptolites from the Valhallfonna Formation, northern Spitsbergen. *Special Papers in Palaeontology 13*, 87–98.

Balashova, E.A. 1964: Morphology, phylogeny and stratigraphic occurrence of the early Ordovician subfamily Ptychopyginae in the Baltic Platform. *Voprosii Paleontologii 4*, 3–56.

Balashova, E.A. 1976: *Systematics of Asaphina (Trilobita) and Their Representatives in the USSR*, 214 pp. Nedra, Leningrad (St Petersburg). [in Russian].

Barton, D.C. 1915: A revision of the Cheirurinae, with notes on their evolution. *Washington University Study of Sciences Series 3*, 101–152.

Begg, J.L. 1945: The illaenid pygidium with special reference to the reflected border. *Geological Magazine 82*, 107–113.

Billings, E. 1859: Fossils of the Calciferous Sandrock, including those of a deposit in white limestone at Mingan, supposed to belong to the formation. *Canadian naturalist and geologist 4*, 345–367.

Billings, E. 1860: On some new species of fossils from the limestone near Point Levis opposite Quebec. *Canadian Naturalist (Ottawa) 5*, 301–324.

Billings, E. 1865: *Palaeozoic Fossils. Volume 1. Containing Descriptions and Figures of New or Little Known Species of Organic Remains from the Silurian Rocks of Canada*, 462 pp. Geological Survey of Canada, Ottawa.

Bockelie, T.G. 1980: Early Ordovician chitinozoa from Spitsbergen. *Palynology 4*, 1–14.

Bockelie, T.G. 1981: Internal morphology of two species of *Lagenochitina* (Chitinozoa). *Review of Palaeobotany and Palynology 34*, 149–164.

Bockelie, T.G. & Fortey, R.A. 1976: An early Ordovician vertebrate. *Nature 260*, 36–38.

Bockelie, T.G., Bruton, D.L. & Fortey, R.A. 1976: Research on the Ordovician rocks of North Ny Friesland, Spitsbergen. *Arbok Norsk Polarinstitutt (for 1975)* 214–217.

Boyce, W.D. 1983: Early Ordovician trilobite faunas of the Boat Harbour and Catoche Formations (St George Group) in the Boat Harbour- Cape Norman area, western Newfoundland. Unpublished MSc thesis, Memorial University of Newfoundland, 272 pp.

Boyce, W.D. 1989: Early Ordovician trilobite faunas of the Boat Harbour and Catoche Formations (St George Group) in the Boat Habour Cape Norman area, Great Northern Peninsula, western Newfoundland. *Report of the Newfoundland Department of Mines and Energy, Geological Survey Branch, 89*, 1–169.

Boyce, W.D. & Knight, I. 2010: Macropalaeontological Investigation of the Upper St George Group, west Isthmus Bay to East Bay section, Port au port Peninsula, western Newfoundland. *Current Research Newfoundland Geological Survey Report 10*, 219–244.

Boyce, W.D. & Stouge, S. 1997: Trilobite and conodont biostratigraphy of the St George Group, Eddies Cove west area, western Newfoundland. *Report of the Newfoundland Department of Mines and Energy, Geological Survey Branch, 97*, 183–200.

Boyce, W.D., McCobb, L.M.E. & Knight, I. 2011: Stratigraphic studies of the Watts Bight Formation (St George Group), Port au port Peninsula, western Newfoundland. *Current Research Report, Newfoundland and Labrador Department of Natural Resources Geological Survey 11*, 215–240.

Bradley, J.H. 1925: Trilobites of the Beekmantown in the Phillipsburg region of Quebec. *Canadian Field Naturalist 39*, 5–9.

Brett, C.E. 1995: Sequence stratigraphy, biostratigraphy and taphonomy in shallow marine environments. *Palaios 10*, 597–616.

Brett, K.D. & Westrop, S.R. 1996: Trilobites of the Lower Ordovician (Ibexian) Fort Cassin Formation, Champlain Valley region, New York State and Vermont. *Journal of Paleontology 70*, 408–427.

Bridge, J. & Cloud, P.E. 1947: New gastropods and trilobites critical in the correlation of Lower Ordovician rocks. *The American Journal of Science 245*, 545–559.

Bruton, D.L. & Owen, A.W. 1988: The Norwegian Upper Ordovician illaenid trilobites. *Norsk Geologisk Tidsskrift 68*, 241–258.

Callaway, C. 1877: On a new area of Upper Cambrian rocks in South Shorpshire with description of a new fauna. *Quarterly Journal of the Geological Society of London 33*, 652–672.

Cocks, L.R.M. & Torsvik, T.H. 2011: The Palaeozoic geography of Laurentia and western Laurussia: A stable craton with mobile margins. *Earth Science Reviews 106*, 1–51.

Cooper, B.N. 1953: Trilobites from the Lower Champlainian formations of the Appalachian Valley. *Memoirs of the Geological Society of America 55*, 1–69.

Cooper, R.A. & Fortey, R.A. 1982: The Ordovician graptolites of Spitsbergen. *Bulletin of the British Museum (Natural History): Geology 36*, 157–302.

Cowie, J.W. & Adams, P.J. 1957: The geology of the Cambro-Ordovician rocks of central east Greenland: Part 1. *Meddelelser om Grønland 153*, 1–193.

Cullison, J.S. 1944: The stratigraphy of some Lower Ordovician formations of the Ozark Uplift. *University of Missouri School of Mines and Metallurgy, Technical Bulletin 15*, 1–112.

Dalman, J.W. 1827: Om Palaeaderna eller de så kallade Trilobiterna. *Kungliga Svenska Vetenskaps-Akadmiens Handlingar 1827*, 1–109.

Dean, W.T. 1989: Trilobites from the SurveyPeak, Outram and Skoki Formations (Upper Cambrian – Lower Ordovician) at Wilcox Pass, Jasper National Park, Alberta, Canada. *Bulletin of the Geological Survey of Canada 389*, 1–141.

Derby, J., Fritz, R., Longacre, S., Morgan, W. & Sternbach, C. (eds) 2012: The Great American Carbonate Bank: The geology and economic resources of the Cambrian-Ordovician Sauk Megasequence of Laurentia. *Memoirs of the American Association of Petroleum Geologists 98*, 528.

Desbiens, S., Bolton, T.E. & McCraken, A.D. 1996: Fauna of the lower Beauharnois Formation (Beekmanown Group, Lower Ordovician) Grande-Ile, Quebec. *Canadian Journal of Earth Sciences 33*, 1132–1153.

Dwight, W.B. 1884: Recent explorations in the Wappinger valley limestones and other formations of Duchess Co., N.Y. No. 4 Descriptions of Calciferous (?) fossils. *American Journal of Science 27*, 249–259.

Ebbestad, J.O.R. 1999: Trilobites of the Bjørkåsholmen formation in the Oslo region, Norway. *Fossils and Strata 47*, 1–118.

Ethington, R.L. & Clark, D.L. 1981: Lower and Middle Ordovician conodonts from the Ibex area, western Millard County, Utah. *Geology Studies, Brigham Young University 28*, 1–155.

Ethington, R.L., Engel, K.M. & Elliott, K.L. 1987: An abrupt change in conodont faunas in the Lower Ordovician of the Midcontinent Province, 111–127. *In* Aldridge, R.J. (ed.): *Paleobiology of Conodonts*. Ellis Horwood, Chichester, UK.

Evans, D.H. 2011: Nautiloids from the Durness Limestone, northwest Scotland. *Palaeontographical Society Monograph 637*, 1–151.

Flower, R.H. 1964: The nautiloid Order Ellesmeroceratida (Caphalopoda). *Memoir of the New Mexico Bureau of Mines and Mineral Resources 12*, 1–234.

Flower, R.H. 1968: Fossils from the Smith Basin Limestone of the Fort Ann Region. *Memoir of the New Mexico Bureau of Mines and Mineral Resources 22*, 23–57.

Fortey, R.A. 1971: *Tristichograptus*, a triserial graptolite from the Lower Ordovician of Spitsbergen. *Palaeontology 14*, 188–199.

Fortey, R.A. 1973: A new pelagic trilobite from the Ordovician of Spitsbergen, Ireland, and Utah. *Palaeontology 17*, 111–124.

Fortey, R.A. 1974: The Ordovician trilobites of Spitsbergen: 1. Olenidae. *Skrifter Norsk Polarinstitutt 160*, 1–81.

Fortey, R.A. 1975a: The Ordovician trilobites of Spitsbergen: 2. Asaphidae, Nileidae, Raphiophoridae and Telephinidae of the Valhallfonna Formation. *Skrifter Norsk Polarinstitutt 162*, 1–125.

Fortey, R.A. 1975b: Early Ordovician trilobite communities. *Fossils & Strata 4*, 331–352.

Fortey, R.A. 1976: Correlation of shelly and graptolitic early Ordovician successions, based on the sequence in Spitsbergen. *In* Bassett, M.G. (ed.): *The Ordovician System*. University of Wales Press and National Museum of Wales, Cardiff, pp. 263–280.

Fortey, R.A. 1979: Early Ordovician trilobites from the Catoche Formation (St George Group) western Newfoundland. *Bulletin of the Geological Survey of Canada 321*, 61–114.

Fortey, R.A. 1980a: The Ordovician trilobites of Spitsbergen: 3. Remaining trilobites of the Valhallfonna Formation. *Skrifter Norsk Polarinstitutt 171*, 1–113.

Fortey, R.A. 1980b: The Ordovician of Spitsbergen and its relevance to the base of the Middle Ordovician in North America. *In* Wones, D.R. (ed.): *The Caledonides in the USA IGCP Project 27*, 33–40. Virginia Tech, Blacksburg, VA.

Fortey, R.A. 1980c: Generic longevity in Lower Ordovician trilobites: Relation to environment. *Paleobiology 6*, 24–31.

Fortey, R.A. 1983: Cambrian-Ordovician trilobites from the boundary beds in Western Newfoundland and their phylogenetic significance. *Special Papers in Palaeontology 30*, 179–211.

Fortey, R.A. 1984: Global Ordovician transgressions and regressions and their biological implications. *In* Bruton, D.L. (ed.): *Aspects of the Ordovician System*, 37–50. Universitetsforlaget, Oslo, 228 pp.

Fortey, R.A. 1986: Early Ordovician trilobites from the Wandel Valley Formation, eastern North Greenland. *Rapport Grønlands Geologiske Undersøgelse 132*, 15–25.

Fortey, R.A. 1988: The Ordovician trilobite *Hadrohybus* Raymond, 1925, and its family relationships. *Postilla 202*, 1–7.

Fortey, R.A. 1990: Ontogeny, hypostome attachment and trilobite classification. *Palaeontology 33*, 529–576.

Fortey, R.A. 1992: Ordovician trilobites from the Durness Group, North-West Scotland, and their palaeobiogeography. *Scottish Journal of Geology 28*, 115–121.

Fortey, R.A. & Barnes, C.R. 1977: Early Ordovician conodont and trilobite communities of Spitsbergen: Influence on biogeography. *Alcheringa 1*, 297–309.

Fortey, R.A. & Bruton, D.L. 1973: Cambrian-Ordovician rocks adjacent to Hinlopenstretet, North Ny Friesland, Spitsbergen. *Bulletin of the Geological Society of America 84*, 2227–2242.

Fortey, R.A. & Cocks, L.R.M. 2003: Palaeontological evidence bearing on Ordovician-Silurian continental reconstructions. *Earth Science Reviews 61*, 245–307.

Fortey, R.A. & Droser, M.L. 1996: Trilobites from the base of the Middle Ordovician, western United States. *Journal of Paleontology 70*, 73–99.

Fortey, R.A. & Droser, M.L. 1999: Trilobites from the base of the type Whiterockian (Middle Ordovician) in Nevada. *Journal of Paleontology 73*, 182–201.

Fortey, R.A. & Holdsworth, B.K. 1971: The oldest known well-preserved radiolaria. *Bolletino della Societa Paleontologica Italiana 10*, 35–41.

Fortey, R.A. & Morris, S.F. 1978: Discovery of nauplius-like trilobite larvae. *Palaeontology 21*, 823–833.

Fortey, R.A. & Owens, R.M. 1975: Proetida, a new Order of trilobites. *Fossils & Strata 4*, 227–239.

Fortey, R.A. & Owens, R.M. 1990: Evolutionary radiations in the Trilobita. *In* Taylor, P.D., Larwood, G.P. (eds): *Major Evolutionary Radiations*. Systematics Association Special, volume 42, 139–164. Clarendon Press, Oxford.

Fortey, R.A. & Owens, R.M. 1991: A trilobite fauna from the highest Shineton Shales in Shropshire, and the correlation of the latest Tremadoc. *Geological Magazine 122*, 437–464.

Fortey, R.A. & Peel, J.S. 1983: The anomalous bathyurid trilobite *Ceratopeltis* and its homoeomorphs. *Special Papers in Palaeontology 30*, 51–57.

Fortey, R.A. & Peel, J.S. 1990: Early Ordovician trilobites from the Poulsen Cliff Formation, Washington Land, western North Greenland. *Bulletin of the Geological Society of Denmark 38*, 11–32.

Fortey, R.A. & Whittaker, J.E.P. 1976: *Janospira* – an Ordovician microfossil in search of a phylum. *Lethaia 9*, 397–403.

Fortey, R.A. & Wilmot, N.V. 1991: Trilobite cuticle thickness in relation to palaeoenvironment. *Palaeontologische Zeitschrift 65*, 141–151.

Gobbett, D.J. & Wilson, C.B. 1960: The Oslobreen Series Upper Hecla Hoek of Ny Friesland, Spitsbergen. *Geological Magazine 97*, 441–457.

Hallam, A. 1958: A Cambro-Ordovician fauna from the Hecla Hoek succession of Ny Friesland, Spitsbergen. *Geological Magazine 95*, 71–76.

Hansen, T. 2009: Trilobites of the Middle Ordovician Elnes Formation of the Oslo Region, Norway. *Fossils & Strata 56*, 1–215.

Hansen, J. & Holmer, L.E. 2010: Diversity fluctuations and biogeography of Ordovician brachiopod faunas in northeastern Spitsbergen. *Bulletin of Geosciences 85*, 497–504.

Hansen, J. & Holmer, L.E. 2011: Taxonomy and biostratigraphy of Ordovician brachiopods from northeastern Ny Friesland, Spitsbergen. *Zootaxa 3076*, 1–122.

Harland, W.B. 1997: The geology of Svalbard. *Memoir of the Geological Society of London 17*, 877.

Hawle, I. & Corda, A.J.C. 1847: Prodrom einer Monographie der böhmischen Trilobiten. *Köngliche böhmischen Gesellschaft der Wissenschaften Abhandlung 5*, 1–176.

Heller, R.L. 1954: Stratigraphy and paleontology of the Roubidoux Formation of Missouri. *Missouri Geological Survey and Water Resources, series 2, 35*, 1–13.

Hintze, L.F. 1953: Lower Ordovician trilobites from western Utah and eastern Nevada. *Bulletin of the Utah Geological and Mineralogical Survey 48*, 1–249.

Hintze, L.F. 1973: Lower and Middle Ordovician stratigraphic sections in the Ibex area, Millard County, Utah. *Geology Studies Brigham Young University 20*, 3–36.

Hintze, L.F. & Jaanusson, V. 1956: Three new genera of asaphid trilobites from the Lower Ordovician of Utah. *Bulletin of the Geological Institution, University of Uppsala 36*, 51–57.

Hoel, O.A. 1999: Trilobites of the Hargastrand Member (Toyen Formation, lowermost Arenig) from the Oslo Region, Norway. Part II: Remaining non-asaphid groups. *Norsk Geologisk Tidsskrift 79*, 259–280.

Holdsworth, B.K. 1977: Paleozoic Radiolaria: Stratigraphic distribution in Atlantic borderlands. *In* Swain, F.M. (ed.): *Stratigraphic Micropalaeontology of Atlantic Basin and Borderlands.* Developments in Paleontology and Stratigraphy 6, 167–184. Elesvier, New York, 599 pp.

Holloway, D.J. 2007: The trilobite *Protostygina* and the composition of the Styginidae, with two new genera. *Paläontologische Zeitschrift 81*, 1–16.

Holm, G. 1882: De svenske artena af Trilobitslaktet *Illaenus* (Dalman). *Kungliga Svenska Vetenskaps-Akademiens Handlingar 7*, 1–148.

Hupé, P. 1953: Classe des Trilobites. *In* Piveteau, J. (ed.): *Traité de Paléontologie. Tome 3*, 344–426. Masson et cie, Paris, 1063 pp.

Hupé, P. 1955: Classification de Trilobites. *Annales de Paléontologie 41*, 91–325.

Jaanusson, V. 1954: Zur Morphologie und Taxonomie der Illaeniden. *Arkiv för mineralogi och geologi 1*, 545–583.

Jaanusson, V. 1957: Underordovizische Illaeniden aus Skandinavien. *Bulletin of the Geological Institutions of the University of Uppsala 37*, 79–165.

Jablonski, D. 2005: Evolutionary innovations in the fossil record: The intersection of ecology, development and macroevolution. *Journal of Experimental Zoology 304B*, 58–77.

Jablonski, D. & Bottjer, D.J. 1990: Onshore-offshore trends in marine invertebrate evolution. *In* Ross, R.M., Allmon, W.D. (eds): *Causes of Evolution: A Paleontological Perspective*, 21–75. University of Chicago Press, 494 pp.

James, N.P., Stevens, R.K., Barnes, C.R. & Knight, I. 1989: Evolution of a lower Paleozoic continental margin carbonate platform, northern Canadian Appalachians. *In* Crevello, P., Sarg, R., Read, J.F., Wilson, J.L. (eds): *Controls on Carbonate Platform Development.* Society of Economic Paleontologists and Mineralogists, Special Publication 44, 123–146.

Ji, J.-L. & Barnes, C.R. 1994: Lower Ordovician conodonts of the St. George Group, Port au Port Peninsula, western Newfoundland, Canada. *Palaeontographica Canadiana 11*, 1–149.

Ji, J.-L. & Barnes, C.R. 1996: Cambrian and Lower Ordovician conodont biostratigraphy of the Survey Peak Formation (Ibexian/Tremadoc), Wilcox Pass, Alberta, Canada. *Journal of Paleontology 70*, 271–290.

Knight, I. & James, N.P. 1988: *Stratigraphy of the Lower to Lower Middle Ordovician St. George Group, western Newfoundland.* Government of Newfoundland and Labrador, Department of Mines, Mineral Development Division, Report 88-4, 48 pages.

Kobayashi, T. 1934: The Cambro-Ordovician formations and faunas of South Chosen. Palaeontology. Part 2. Lower Ordovician faunas. *Journal of the Faculty of Science Imperial University of Tokyo 3*, 521–585.

Kobayashi, T. 1935: The Cambro-Ordovician formations and faunas of South Chosen: III. Cambrian faunas of South Chosen with special study of the Cambrian trilobite genera and families. *Journal of the Faculty of Science Imperial University of Tokyo 4*, 49–344.

Kobayashi, T. 1940: Lower Ordovician faunas from Caroline Creek near Latrobe, Mersey River District, Tasmania. *Proceedings of the Royal Society of Tasmania for 1939*, 67–76.

Kobayashi, T. 1955: The Ordovician fossils of the McKay Group in British Columbia, western Canada, with a note on the Ordovician palaeogeography. *Journal of the Faculty of Science, Imperial University of Tokyo, Section 2*, 9, 355–493.

Kosteva, N.N. & Teben'kov, A.M. 2006: Lithological description of Cambrian-Ordovician deposits of Hinlopenstretet, Spitsbergen. *Complex Investigations of Spitsbergen's Nature*, Issue 6, 109–119. Publ. RSC RAS, Apatity. [in Russian].

Lake, P. 1907: British Cambrian trilobites. *Monographs of the Palaeontographical Society 296*, 27–89.

Landing, E., Adrain, J.M., Westrop, S.R. & Kröger, B. 2012: Tribes Hill- Rochdale Formations in eastern Laurentia: Proxies for early Ordovician (Tremadocian) eustasy on a tropical passive margin (New York and west Vermont). *Geological Magazine 149*, 93–123.

Lehnert, O., Stouge, S. & Brandl, P. 2013: Conodont biostratigraphy in the Early to Middle Ordovician strata of the Oslobreen Group in Ny Friesland, Svalbard. *Zeitskrift. Dt. Ges. Geowissenscaften 164*, 149–172, Stuttgart.

Loch, J.D. 2007: Trilobite biostratigraphy and correlation of the Kindblade Formation (Lower Ordovician) of Carter and Kiowa Counties, Oklahoma. *Bulletin of the Oklahoma Geological Survey 149*, 1–154, 28 pls.

Lochman, C. 1966: Lower Ordovician (Arenig) faunas from the Williston Basin of Montana and North Dakota. *Journal of Paleontology 40*, 512–548.

Ludvigsen, R. 1979: Lower Ordovician trilobites of the Oxford Formation, eastern Ontario. *Canadian Journal of Earth Sciences 16*, 859–865.

Ludvigsen, R., Westrop, S.R. & Kindle, C.H. 1989: Sunwaptan (Upper Cambrian) Trilobites of the Cow Head Group, western Newfoundland. *Palaeontographica Canadiana 6*, 1–175.

Maletz, J. & Bruton, D.L. 2007: Lower Ordovician (Chewtonian to Castelmanian) radiolarians of Spitsbergen. *Journal of Systematic Palaeontology 5*, 245–288.

Maletz, J. & Bruton, D.L. 2008: The Middle Ordovician *Provenocitum procerulum* radiolarian assemblage of Spitsbergen and its biostratigraphic correlation. *Palaeontology 51*, 1181–1200.

Marek, L. 1952: Contribution to the stratigraphy and faunas of the upper part of the Králův Dvůr Shales (Ashgillian). *Sbornik Ústředního Ústavu Geologického 19*, 429–455.

McCobb, L.M.E. & Owens, R.M. 2008. Ordovician (Ibex-Whiterock) trilobites from central East Greenland. *In* Rabano, I., Gozalo, R., Garcia-Bellido, D. (eds): *Advances in Trilobite Research*, 253–258. Cuadernos del Museo Geominero 9, 1–448.

McCobb, L.M.E., Boyce, W.D. & Knight, I. 2011: Correlation of Lower Ordovician (Ibexian) faunas in North Eastern Greenland and western Newfoundland – new trilobite and lithostratigraphic data. *In* Guttierez Marco, J.C., Rabano, I., Garcia Bellido, D. (eds): *Ordovician of the World*, 339–346, Madrid.

McCobb, L.M.E., Boyce, W.D., Knight, I. & Stouge, S. in press: Early Ordovician trilobites from the Antiklinalbugt Formation, North-east Greenland. *Journal of Paleontology* (in press).

McCormick, T. & Fortey, R.A. 2002: The Ordovician trilobite *Carolinites*, a test case for microevolution in a macrofossil lineage. *Palaeontology 45*, 229–257.

Mergl, M. 2006: Tremadocian trilobites of the Prague Basin, Czech Republic. *Acta Musei Nationalis Pragae, series B 62*, 1–70.

Moore, R.C. (ed.) 1959: *Treatise on Invertebrate Paleontology*. Part O. Arthropoda 1. The Geological Society of America and University of Kansas Press, New York City and Lawrence, Kansas, 560 pp.

Morris, N.J. & Fortey, R.A. 1976: *Tironucula* gen. nov. and its significance in bivalve evolution. *Journal of Paleontology 50*, 701–709.

Nielsen, A.T. 1995: Trilobite systematics, biostratigraphy and palaeoecology of the Lower Ordovician Komstad Limestone and Huk Formations, southern Scandinavia. *Fossils & Strata 38*, 374.

Nielsen, A.T. 2004: Ordovician sea level changes: A Baltoscandian perspective. *In* Webby, B.D.E., Paris, F., Droser, M., Percival, I.G. (eds): *The Great Ordovician Biodiversification Event*, 84–93. Columbia University Press, New York.

Owen, A.W. & Bruton, D.L. 1980: Late Caradoc-early Ashgill trilobites of the central Oslo Region, Norway. *Paleontological Contributions from the University of Oslo No. 245*, 163.

Poulsen, C. 1927: The Cambrian, Ozarkian and Canadian faunas of Northeast Greenland. *Meddelelser om Grønland 70*, 233–343.

Poulsen, C. 1937: On the Lower Ordovician faunas of east Greenland. *Meddelelser om Grønland 119*, 1–72.

Poulsen, C. 1946: Notes on Cambro-Ordovician fossils collected by the Oxford University Ellesmere Land Expedition, 1934–1935. *Quarterly Journal of the Geological Society of London 102*, 299–337.

Poulsen, C. 1948: *Grinnellaspis*, a new name replacing *Actinopeltis* Poulsen, 1946. *Journal of Paleontology 22*, 107.

Poulsen, C. 1951: The position of the East Greenland Cambro-Ordovician in the palaeogeography of the North Atlantic region. *Meddelelser Dansk Geologisk Forening 12*, 161–162.

Poulsen, V. 1965: An early Ordovician trilobite fauna from Bornholm. *Meddelelser fra Dansk Geologisk Forening 16*, 49–113.

Poulsen, V. 1969: The types of *Raymondaspis limbata* (Angelin, 1854) Class Trilobita. *Geologiska Föreningens I Stockholm Förhandlingar 91*, 407–416.

Prantl, F. & Přibyl, A. 1949: O novych nebo málo známych trilobitech ceského ordoviku. *Rozpravy Ceske Akademie Ved a Umeni 58*, 1–22.

Pratt, B.J. & James, N.P. 1986: The St. George Group (Lower Ordovician) of western Newfoundland: Tidal flat island model for carbonate sedimentation in epeiric seas. *Sedimentology 33*, 313–343.

Pratt, B.J. & James, N.P. 1989: Early Ordovician thrombolite reefs, St. George Group, western Newfoundland. *In* Geldsetzer, H.H.J., James, N.P., Tebbutt, G.E. (eds): *Reefs, Canada and Adjacent Areas.* Memoir of the Canadian Society of Petroleum Geologists 13, 231–240.

Raymond, P.E. 1905: The trilobites of the Chazy Limestone. *Annals of the Carnegie Museum 3*, 328–386.

Raymond, P.E. 1913: A revision of the species which have been referred to the genus *Bathyurus*. *Bulletin of the Victoria Memorial Museum 1*, 51–69.

Raymond, P.E. 1925: Some trilobites of the lower Middle Ordovician of eastern North America. *Bulletin of the Museum of Comparative Zoology, Harvard University 67*, 1–180.

Repetski, J.E. 1982: Condonts of the El Paso Group (Lower Ordovician) of westernmost Texas, and southern New Mexico. *Memoir of the New Mexico Bureau of Mines and Mineral Resources 40*, 1–121.

Riding, R. & Toomey, D.F. 1972: The sedimentological role of *Epiphyton* and *Renalcis* in lower Ordovician mounds, southern Oklahoma. *Journal of Paleontology 46*, 509–519.

Rohr, D.M., Measures, E.A., Boyce, W.D. & Knight, I. 2001: Early Ordovician gastropods of the Barbace Cove Member (Boat Harbour Formation) and Catoche Formation, western Newfoundland. *Current Research, Newfoundland Department of Mines and Energy, Geological Survey Report 2001 1*, 113–126.

Ross, R.J. 1951: Stratigraphy of the Garden City Formation in northeastern Utah and its trilobite fauna. *Bulletin of the Peabody Museum of Natural History 6*, 1–161.

Ross, R.J. 1953: Additional Garden City (early Ordovician) trilobites. *Journal of Paleontology 27*, 633–646.

Ross, R.J. 1967: Some Middle Ordovician brachiopods and Trilobites from the Basin Ranges, western United States. *Professional Papers of the U.S. Geological Survey 523D*, 1–43.

Ross, R.J. 1970: Ordovician brachiopods, trilobites and stratigraphy in eastern and central Nevada. *Professional papers of the United States Geological Survey 639*, 1–103.

Ross, R.J., Adler, F.J., Amsden, T.W. *et al.* 1982: The Ordovician System in the United States –correlation chart and explanatory notes. *Publication of the International Union of Geological Sciences 12*, 1–73.

Ross, R.J., Hintze, L.F., Ethington, R.L., Miller, J.F., Taylor, M.E. & Repetski, J.E. 1997: The Ibexian Series in the North American Ordovician. *Professional Papers of the United States Geological Survey 1579-A*, 1–50.

Salter, J.W. 1864: A monograph of the British trilobites, part 1. *Monographs of the Palaeontographical Society, 16*, 1–80.

Sdzuy, K. 1955: Die Fauna der Liemitz Schiefer (Tremadoc). *Abhandlungen der Senckenbergischen Naturforschenden Gesellschaft 492*, 1–74.

Sepkoski, J.J. 1991: A model of onshore-offshore change in faunal diversity. *Paleobiology 17*, 58–77.

Shaw, F.C. & Bolton, T. 2011: Ordovician Trilobites from the Romaine and Mingan Formations (Ibexian-Late Whiterockian), Mingan Islands, Quebec. *Journal of Paleontology 85*, 406–411.

Smith, M.P. 1991: Early Ordovician condonts of East and North Greenland. *Meddelelser om Groenland 26*, 1–81.

Smith, M.P. & Bjerreskov, M. 1992: *The Ordovician System in Greenland*. International Union of Geological Sciences, Special publication 29A.

Smith, M.P. & Rasmussen, J.A. 2008: Cambrian-Silurian development of the Laurentian margin of the Iapetus Ocean in Greenland and related areas. *Memoir of the Geological Society of America 202*, 137–167.

Stouge, S. & Bagnoli, G. 1988: Early Ordovician conodonts from Cow Head, western Newfoundland. *Palaeontographica Italiana 75*, 89–178.

Stouge, S., Christiansen, J.L. & Holmer, L.E. 2011: Lower Palaeozoic stratigraphy of Murchisonfjorden and Sparreneset, Nordaustlandet, Svalbard. *Geografiska Annalen 93*, 209–226.

Stouge, S., Boyce, W.D., Christiansen, J.L., Harper, D.A.T. & Knight, I. 2012: Development of the lower Cambrian–Middle Ordovician carbonate platform: North Atlantic Region, 307a–312a, 597–626. *In* Derby, J., Fritz, R., Longacre, S., Morgan, W., Sternbach, C. (eds): *The Great American Carbonate Bank. The geology and economic resources of the Cambrian-Ordovician Sauk Megasequence of Laurentia*. Memoirs of the American Association of Petroleum Geologists 98, 528 pp.

Sweet, W. & Tolbert, C.M. 1997: An Ibexian (Lower Ordovician) reference section in the southern Egan Range, Nevada, for a conodont based Chronostratigraphy. *Professional papers of the U.S. Geological Survey 1579-B*, 53–84.

Swett, K. 1981: Cambro–Ordovician strata in Ny Friesland, Spitsbergen, and their palaeotectonic significance. *Geological Magazine 118*, 225–250.

Swett, K. & Smit, D.E. 1972: Palaeogeography and depositional environments of the Cambro-Ordovician shallow marine facies of the North Atlantic. *Bulletin of the Geological Society of America 83*, 3223–3248.

Taylor, J.F., Repetski, J.E., Loch, J.D. & Leslie, S.A. 2012: Biostratigraphy and chronostratigraphy of the Cambrian-Ordovician Great American Carbonate Bank, 15–35. *In* Derby, J., Fritz, R., Longacre, S., Morgan, W., Sternbach, C. (eds): *The Great American Carbonate Bank. The geology and economic resources of the Cambrian-Ordovician Sauk Megasequence of Laurentia*. Memoirs of the American Association of Petroleum Geologists 98, 528 pp.

Thompson, T.L. 1991: Paleozoic succession in Missouri, part 2, Ordovician. *Report of Investigation Missouri Department of Natural Resources 70*, 1–282.

Tjernvik, T.E. 1956: On the early Ordovician of Sweden. Stratigraphy and fauna. *Bulletin of the Geological Institute, University of Uppsala 36*, 107–284.

Toomey, D.F. & Nitecki, M.H. 1979: Organic build-ups in the Lower Ordovician (Canadian) of Texas and Oklahoma. *Fieldiana (Geology) n.s. 2*, 1–181.

Twenhofel, W.H. 1938: Geology and paleontology of the Mingan Islands, Quebec. *Special Papers of the Geological Society of America, 11*, 1–132.

Vallance, G. & Fortey, R.A. 1968: Ordovician succession in North Spitsbergen. *Proceedings of the Geological Society of London 1648*, 91–97.

Vogdes, A.W. 1890: A bibliography of Palaeozoic Crustacea from 1698 to 1899, including a list of North American species and a systematic arrangement of genera. *Bulletin of the U.S. Geological Survey 63*, 1–177.

Waisfeld, B.G., Vaccari, N.G., Chatterton, B.D.E. & Edgecombe, G.E. 2001: Systematics of Shumardiidae (Trilobita) with new species from the Ordovician of Argentina. *Journal of Paleontology 75*, 827–854.

Walcott, C.D. 1886: Second contribution to the studies on the Cambrian faunas of North America. *Bulletin of the U.S. Geological Survey 30*, 1–369.

Walcott, C.D. 1924: Cambrian and Lower Ozarkian trilobites. *Smithsonian Miscellaneous Collections 75*, 53–60.

Webby, B.D.E., Paris, F., Droser, M. & Percival, I.D. (eds) 2004: *The Great Ordovician Biodiversification Event*. Columbia University Press, New York, 484 pp.

Weller, S. & St Clair, S. 1928: Geology of Ste. Genevieve County, Missouri. *Bulletin of the Missouri Bureau of Geology and Mines 22*, 1–352.

Whitfield, R.P. 1886: Notice of the geological investigations along the eastern shore of Lake Champlain, conducted by Prof H. M. Seely and Prest. Ezra Brainerd of Middlebury College, with descriptions of the new fossils discovered. *Bulletin of the American Museum of Natural History 1*, 293–345.

Whitfield, R.P. 1889: Observations on some imperfectly known fossils from the Calciferous sandrock of Lake Champlain, and descriptions of several new forms. *Bulletin of the American Museum of Natural History 2*, 41–63.

Whitfield, R.P. 1890: Observations of the fauna of the rock at Fort Cassin, Vermont, with descriptions of a few new species. *Bulletin of the American Museum of Natural History 3*, 25–39.

Whitfield, R.P. 1897: Descriptions of new species of Silurian Fossils from near Fort Cassin and elsewhere on Lake Champlain. *Bulletin of the American Museum of Natural History 9*, 177–184.

Whittington, H.B. 1953: North American Bathyuridae and Leiostegiidae. *Journal of Paleontology 27*, 647–678.

Whittington, H.B. 1963: Middle Ordovician trilobites from Lower Head, western Newfoundland. *Bulletin of the Museum of Comparative Zoology, Harvard University 129*, 1–118.

Whittington, H.B. 1965: Trilobites of the Ordovician Table Head Formation, Western Newfoundland. *Bulletin of the Museum of Comparative Zoology, Harvard University 132*, 275–442.

Whittington, H.B. 1988: Hypostomes of post-Cambrian trilobites. *Memoir of New Mexico Bureau of Mines and Mineral Resources 44*, 321–339.

Whittington, H.B. & Kindle, C.H. 1969: Cambrian and Ordovician stratigraphy of western Newfoundland. *Memoir of the American Association of Petroleum Geologists 12*, 655–664.

Whittington, H.B., Chatterton, B.D.E., Speyer, S.E., Fortey, R.A., Owens, R.M., Chang, W.-T., Dean, W.T., Jell, P.A., Laurie, J.R., Palmer, A.R., Repina, L.N., Rushton, A.W.A., Shergold, J.H., Clarkson, E.N.K., Wilmott, N.V. & Kelly, S.R.A. 1997: *Treatise on Invertebrate Paleontology, Part O, Arthropoda 1, Trilobita (revised). Volume 1*. The Geological Society of America and University of Kansas Press, New York and Lawrence, Kansas.

Williams, M. & Siveter, D.J. 2008: The earliest leperditicope arthropod: A new genus from the Ordovician of Spitsbergen. *Journal of Micropalaeontology 27*, 97–101.

Wiman, C. 1905: Ein Shumardiascheifer bei Lanna in Nerike. *Arkiv för Zoologi 2*, 1–20.

Yochelson, E.L. & Bridge, J. 1957: The Lower Ordovician gastropod *Ceratopea*. *Professional Papers of the U.S. Geological Survey 294-H, 28*, 1–304.

Yochelson, E.L. & Copeland, M.J. 1974: Taphonomy and taxonomy of the early Ordovician gastropod *Ceratopea canadensis* (Billings, 1865). *Canadian Journal of Earth Sciences 11*, 189–207.